# Wireless Charging Technology and the Future of Electric Transportation

# Other SAE books of interest:

**Unwinding Electric Motors: Strategic Perspectives and Insights for Automotive Powertrain Applications**
By Timothy George Thoppil
(Product Code: JPF-P3-003)

**Advanced Hybrid Powertrains for Commercial Vehicles**
By Haoran Hu, Simon Baseley, and Rudolf M. Smaling
(Product Code: R-396)

**Automotive 2030 – North America**
By Bruce Morey
(Product Code: T-127)

# Wireless Charging Technology and the Future of Electric Transportation

Lead Author:
In-Soo Suh (Professor, Korea Advanced Institute of Science and Technology)

Co-Authors:
Dong-Ho Cho (Professor, Korea Advanced Institute of Science and Technology)
Jorg Franke (Professor, FAU Erlangen-Nuremberg)
Soon-Man Hong (Invited Professor, Korea Advanced Institute of Science and Technology)
Sung-Kwan Jung (Research Associate Professor, Korea Advanced Institute of Science and Technology)
Byung-Song Lee (Korea Railway Research Institute)
John M. Miller (Retired, Oak Ridge National Laboratory)
Florian Risch (Research Assistant, FAU Erlangen-Nuremberg)
Naoki Shinohara (Professor, Kyoto University)
Faical Turki (Dipl.-Ing., Paul Vahle GmbH & Co. KG)

SAE INTERNATIONAL®
Warrendale, Pennsylvania
USA

| | |
|---|---|
| SAE INTERNATIONAL | 400 Commonwealth Drive<br>Warrendale, PA 15096-0001 USA<br>E-mail: CustomerService@sae.org<br>Phone: 877-606-7323 (inside USA and Canada)<br>724-776-4970 (outside USA)<br>Fax: 724-776-0790 |



**SAE Order Number R-444**
**DOI 10.4271/R-444**

**Library of Congress Cataloging-in-Publication Data**
Wireless charging technology and the future of electric transportation / chief editor, In-Soo Suh (Professor, Korea Advanced Institute of Science and Technology (KAIST)) ; co-authors, In-Soo Suh (Professor, Korea Advanced Institute of Science and Technology), John M. Miller (Retired, Oak Ridge National Laboratory), Jorg Franke (Professor, FAU Erlangen-Nuremberg), Florian Risch (Research Assistant, FAU Erlangen-Nuremberg), Naoki Shinohara (Professor, Kyoto University), Faical Turki (Dipl.-Ing., Paul Vahle GmbH & Co. KG), Dong Ho Cho (Professor, Korea Advanced Institute of Science and Technology), Soon Man Hong (President, Korea Railroad Research Institute), Sung -Kwan Jung (Research Associate Professor, Korea Advanced Institute of Science and Technology).

pages cm

"Nikola Tesla's Wardenclyffe Tower experiment in Long Island in the early 1900s, the transmittance of electricity from a source to an electric device without a wire or cord has been pursued for more than a century. While current automotive industry is focused on developing static wireless charging of electric vehicles for plug-in hybrid or electric vehicles with the delivering capacity below 7 kW ranges, a few prototype technologies are demonstrated with dynamic wireless charging capabilities with more than several 10 kW ranges. One example described in this book is the historic public launch of a people mover vehicle in Seoul Grand Park, Korea, with dynamic wireless charging capability, which was led by Dr. In-Soo Suh, the principal editor of this book. Around the world, the major automakers are developing their strategies for conductive and wireless charging technologies, with concerted efforts to establish technical standards on wireless electric vehicle charging, mainly focused on the safety considerations and inter-operability"-- Provided by publisher.

Includes bibliographical references.

ISBN 978-0-7680-8153-4

1. Electric vehicles--Batteries. 2. Battery chargers. 3. Electricity in transportation. 4. Electronics in transportation. 5. Electromagnetic induction. 6. Electric power transmission. I. So, In-su, editor. II. Society of Automotive Engineers, issuing body.

TL220.W568 2015
629.25'02--dc23

2015005726



**ISBN-Print 978-0-7680-8153-4**
**ISBN-PDF 978-0-7680-8177-0**
**ISBN-epub 978-0-7680-8179-4**
**ISBN-prc 978-0-7680-8178-7**

**To purchase bulk quantities, please contact:**
SAE Customer Service
E-mail: CustomerService@sae.org
Phone: +1.877.606.7323 (inside USA and Canada)
+1.724.776.4970 (outside USA)
Fax: +1.724.776.0790

# Executive Summary

Beginning with Nikola Tesla's Wardenclyffe Tower experiment in Long Island in the early 1900s, the transmittance of electricity from a source to an electric device without a wire or cord has been pursued for more than a century. Recent research and development outcomes and products in the markets are very encouraging. For example, with the interface standard for inductive electrical power transfer over short distances published, called Qi (pronounced "Chee"), wireless charging for mobile devices gained strong momentum and popularity in the mainstream after the Wireless Power Consortium (WPC) published the Qi low-power transmission specification in August of 2009. Many documents and papers applying wireless power transfer (WPT) to electric vehicles, mobile devices, home and industrial electronics, and medical implants have been published over the last decade along with significant technological improvements, and detailed procedural adoption strategies and regulations toward commercialization.

When KAIST (Korea Advanced Institute of Science and Technology, www.kaist.ac.kr) initiated a research and development project on dynamic wireless charging, called OLEV, in the summer of 2009, numerous questions existed about the potential attraction of the technology as well as achieving the key objectives of transmission distance, efficiency, and other considerations in the commercializing phase in Korea. In those days, electric vehicles were not prominent in market positioning, mainly due to the limited battery technology, insufficient charging infrastructure, concerns on the overall electrical energy supply/demand, and relatively slower changes to regulatory and social resistance, while governmental efforts and global movement toward future green transportation were strong.

Demonstrating several electric buses and trams with the OLEV dynamic wireless charging technology in public places drew strong attention from the global media from 2009–2011, as it addressed the fully electric operating vehicles, on-road charger facilities, and integrated operational parameters such as the system transmission efficiency, battery charging rate, economic analysis, electromagnetic field (EMF) restriction, and further maintenance considerations. As a chief engineer responsible for the historic launch of the OLEV tram in Seoul Grand Park in January 2010, I am so proud of the first round trip of the OLEV tram of 2.3 km without a wired charger. As of 2015, there are four people mover pure electric vehicles in operation every 15 minutes, with dynamic wireless charging as a proven commercial operation in history. In addition, two buses are on service as a part of public city transportation every day in Gumi City, in Korea, as a practical commercialization test-bed. In addition, in KAIST campus, two shuttle buses are fully operating every 15 min with stationary and dynamic wireless charging. It took less than five years starting from the research and development phase to the commercialization of the innovative technology.

*Time* magazine and the World Economic Forum acknowledged KAIST's achievement by selecting it as one of the top innovative transportation technologies in 2010 and 2013, respectively. Along with KAIST's activities, several leading contributors should be noted, such as Oakridge National Laboratory (ORNL), Automotive OEMs, Halo-IPT, MIT, Bombardier, European Consortium-FABRIC, and the recent proposal of British Transportation Research Laboratory (TRL), etc. through the conferences, test-beds, and public announcements.

The recent advancements, starting from late 2010, in establishing a technical standard in wireless charging for electric vehicles, especially SAE J2954, Wireless Charging of Electric & Plug-in Hybrid Vehicles, have been impressive. In general, the technical standard activities in ISO and IEC and other national standard bodies on wireless power transfer have been broad, systematic, and detailed enough to have electric vehicles in public markets during the last several years: for example, IEC 61980, ISO 19363, and IEC CISPER 62764.

This book, *Wireless Charging Technology and the Future of Electric Transportation*, is intended to serve as a reference to describe the current research development status and focus in applying WPT to electric vehicle charging and the future green transportation field. Rather than focusing on a specific technology or technological discussion, a general overview of current research and development activities around the globe are covered in this book. Numerous papers have been published on the technical advancements of wireless power transfer technology after we planned out this book's work scope. The technical overview on those papers, including recent technical advancements, will likely be covered in a separate book or review paper.

In the Introduction chapter, the basic principle of wireless power transfer technology is discussed in relation to future sustainable mobility. In chapter 2, the current technology and market status and projection of electric vehicles are discussed, including a technical feasibility study to apply WPT to electric vehicles. The technological comparison between conductive and wireless charging of electric vehicles is discussed in chapter 3. In chapter 4, a description of the OLEV system and project are given along with technical details of the system. Technical issues and challenges in the developing and adoption phase are discussed in chapter 5, such as communication protocols between transmitters and receivers during the charging process, technical performance parameters, related safety, and EMF emissions. The market considerations together with the strategy from stakeholders and technical standard activities are discussed in chapter 6. In the remaining four chapters, various applications of WPT to consumer electronics, rail, aviation, marine, and off-road transportation are discussed in addition to long-distance electrical energy transfer.

# Table of Contents

# Foreword

In January 2011, I edited the SAE Publication, *Green Technologies and the Mobility Industry*. It addressed the new "green paradigm," which considers the environment as a vital aspect of doing business. The environment and climate change, the price of oil and its security, and economic concerns are all challenges facing regional and global economies. Advances in electric vehicle development are a great example of how the mobility industry is responding with technological solutions that are less polluting, more sustainable, and efficient in the long run. Dr. In-Soo Suh and his colleagues in this publication, *Wireless Charging Technology and the Future of Electric Transportation*, continue this theme and advance the case for electrification as a mobility solution.

Although no industry can ever turn its back to cost efficiencies, the equation now also includes the monetary and nonmonetary price of environmental damage, the lasting impact of the carbon footprint, and the value of being socially responsible. Around the world, governments are increasingly regulating and restricting greenhouse gas (GHG) emissions, and incentivizing the creation and use of nonpolluting technologies.

Over the next decade or so, industry forecasts show increased growth in production of electric vehicles in Asia, despite the low penetration rates thus far. Most analysts are conservative showing penetration rates under 5%, but some of the industry predictions actually foresee up to 15% market share by 2025. More vehicle development programs will continue to take place.

Around the world, the major automakers are developing their strategies for conductive and wireless charging technologies, with concerted efforts to establish technical standards on wireless electric vehicle charging, mainly focused on the safety considerations and interoperability. The operating frequency allocation and electromagnetic field exposure to the human body, in addition to existing electrical safety requirements, have been significant issues for technical standard communities.

While the current automotive industry is focused on developing static wireless charging for plug-in hybrid or electric vehicles, delivering capacity below the 7 kW range, a few prototype technologies have demonstrated dynamic wireless charging capabilities with ranges over 10 kW. One example described in this book is the public launch of a people mover vehicle in Seoul Grand Park, Korea, which was led by Dr. In-Soo Suh.

*Wireless Charging Technology and the Future of Electric Transportation* covers the current status of wireless power transfer (WPT) technology and its potential applications to future road and rail transportation systems. It is a reference and provides an in-depth analysis of the most important areas of interest in this new field, including the following topics:

- Working principles of wireless power transfer technology
- Current technology and its projected future impact on electric vehicles

- Comparison between conductive and wireless charging of electric vehicles
- Introduction to dynamic wireless charging systems
- Technological challenges and international technical standards activities
- Applications in consumer electronics, rail, aviation, marine, and off-road transportation
- Long-distance electrical energy transfer

By providing status on recent research and development, it is an invaluable aid to readers, young engineers, researchers, and others seeking advanced knowledge on these subjects.

Andrew Brown, Jr., PhD, PE, FESD, FSAE, NAE
Vice President & Chief Technologist (retired)
Delphi Automotive
2010 SAE President

# Acknowledgments

As the lead author of this book, I sincerely appreciate the acceptance to my invitation and contributions from the authors. Dr. John Miller, recently retired from ORNL, has been a great teacher and reviewer of this book with valuable feedback and comments. I appreciate the contribution from Professor J. Franke and Dr. Risch from FAU Erlangen-Nuremberg, Germany, who are leading the International Electric Drives Production Conference. I would also like to thank Professor Naoki Shinohara, Kyoto University, Japan, for his enthusiasm and never-ending curiosity and for encouraging me to continue this publication. Dr. Turki from Vahle, Germany, contributed commercialization perspective on WPT application. I also appreciate the valuable contributions of my colleagues at KAIST, Professor D. H. Cho and Research Professor S. Jung. I thank Dr. S. Hong, who recently stepped down from the presidential position of Korean Railway Research Institute (KRRI), for his contribution to the railway application of WPT. Importantly, I deeply appreciate all the members of KAIST OLEV Project from 2009, as we have undergone many challenges together.

Special thanks go out to Dr. Andrew Brown for kind encouragement and thorough feedback through the review process. During the editing process, my PhD student, J. Kim, provided great assistance, and Ms. Monica Nogueria of SAE International provided continuous advice and helpful revisions throughout the drafting process.

In-Soo Suh, PhD, Lead Author
Associate Professor, Graduate School for Green Transportation, KAIST
Associate Editor of IEEE ITS Magazine
Research Planner of Korean National Research Foundation

# Chapter 1
# Introduction

## 1.1 Introduction to Wireless Power Transfer Technology

Wireless power transfer (WPT) is the transmission of electrical energy from a power source to an electrical load without conductors or cables. WPT systems encompass the following four types: acoustic, light, capacitive, and inductive [1-1].

Acoustic WPT is a technology that transmits electrical energy via a power electronic circuit, which consists of a transmitting transducer, into an acoustic wave that is transferred through a medium. Acoustic WPT (also called ultrasonic energy transfer) is applied to energizing implanted biomedical devices at a power level of up to a few 100 mW and has an overall peak power transfer efficiency of less than 40%, typically.

Light WPT transmits electrical energy via a light or optical system that consists of laser diodes and photovoltaic diodes. Although the light WPT technology is similar to the microwave WPT in terms of the transmitting principle, differences in applicability exist. Briefly described, light WPT has less defined electromagnetic force (EMF), but it cannot be utilized for delivering massive amounts of electrical energy due to diffraction losses, as can be done with microwave WPT. The efficiency of light WPT is currently reported as about 20–30%. It is mainly applied in spacecraft platforms and terrestrial technologies [1-2], [1-3], [1-4].

A capacitive WPT system is a technology that transmits electrical energy via two primary metal plates connected through a high-frequency resonant power electronic converter. When two isolated secondary plates are placed in a circuit, an alternating electric field is formed between both sets of plates, resulting in a displacement current. As a result, power can be transferred to a load without direct electrical contact, and a certain freedom of movement between the primary and secondary plates is possible. The power level and efficiency of capacitive WPT systems are in the range of 5–50 W and 50–80%, respectively. Typical applications of capacitive WPT systems include supplying power to light-emitting diode (LED) lamps, mobile phones, soccer-playing robots, sensors for respiratory machines, bio-potential measurement systems, and other devices [1-5], [1-6].

Inductive WPT is a technology in which magnetic fields generated between the primary and secondary coils are coupled with current, without the need for either wires or direct electrical contact with metal conductive plates. The concept is similar to that of power transformers [1-7]. The main advantage of inductive WPT is that the currently available technology can transfer electrical power at greater than 98% efficiency. However, inductive WPT is only effective at a short distance between the primary and secondary coils. In addition, misalignment between the two coils can reduce efficiency.

To resolve the limitations of purely inductive WPT, various approaches of magnetic resonant WPT for increasing the air-gap between two coils have been researched and introduced recently [1-8]. These approaches will be discussed in later chapters. The principle of magnetic resonant WPT is that resonant magnetic coupling occurs when the frequencies of the primary and secondary coils are in a resonant condition. The advantages of magnetic resonant WPT are fewer alignment issues and the possibility to transfer power over longer distance between coils when compared to purely inductive WPT. Resonance-based WPT is applied to various fields such as electronics, medical implants, and transport.

### 1.1.1 Theory of Wireless Power Transfer

Typical conductive power transfer happens on a closed circuit through direct power flow across a conductor. Wireless power transfer can be regarded as a contactless power transfer, which includes inductive or capacitive power transfer such as traditional transformers and capacitors. It is noted that electrical power transmission can be achieved from the primary to the secondary coil of a transformer through a narrow air gap at low frequencies of 50 Hz or 60 Hz. This phenomenon is called a strongly coupled near-field inductive power transfer, or closely coupled contactless power transfer. If the operating frequency is high enough, the inductive coupling between two circuits will strengthen due to the rapidly changing rate of the magnetic field that affects the power transfer across the air gap.

However, for applications requiring higher power and longer distances between the coils, an improvement of power transfer efficiency (i.e., maximizing power factor) is necessary. To improve the power transfer efficiency, resonant capacitors can be used in the circuit so that the resonant frequency of the secondary coil can be tuned with the operating frequency of the primary circuit. When the maximum power factor is achieved, the transfer efficiency can be maximized—this is called a resonance condition.

To explain WPT with resonance, two separate electrical circuits that contain resistance, capacitance, and inductance could be used [1-9]. In terms of description, a schematic diagram can be considered as the equivalent primary and secondary circuits of loosely coupled circuits, as shown in Figure 1.1. Here, the subscripts "1" and "2" refer to the "primary" and "secondary" coil values, respectively, of inductance $L$, resistance $R$, and capacitance $C$. The source voltage of the primary circuit is $V_S(=V_1)$, while $R_l$ represents the load resistance of the secondary circuit. The winding ratio of the primary and secondary coils is referred to as $n$. When the operating angular frequency of the power source is $\omega$, the mutual inductance $M$ can be determined by Equation 1.1.

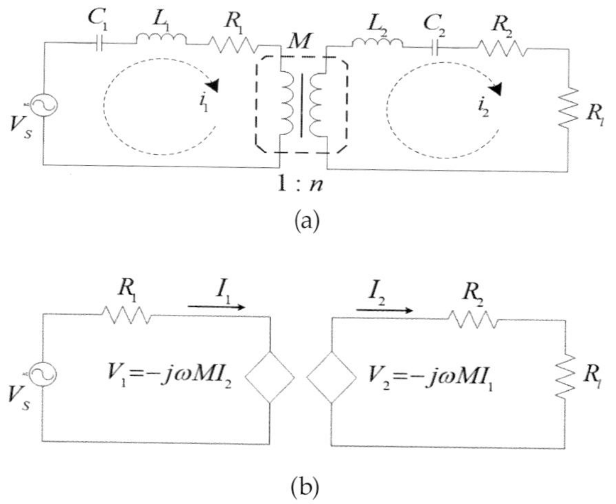


**Figure 1.1** The equivalent circuit of a loosely coupled WPT system.

$$M = \frac{V_S}{\omega I_1} \tag{1.1}$$

By introducing the degree of coupling between the two coils with the coupling coefficient $k$, the mutual inductance can be expressed as Equation 1.2.

$$M = k\sqrt{L_1 L_2} \tag{1.2}$$

where $k$ is the magnetic coupling coefficient indicated by $k = M/\sqrt{L_1 L_2}$. Thus, the circuits in Figure 1.1(a) can be redrawn as in Figure 1.1(b), using mutual inductance.

$$V_1 = \left(\frac{1}{j\omega C_1} + j\omega L_1 + R_1\right) I_1 - j\omega M I_2 \tag{1.3}$$

$$V_2 = j\omega M I_1 - \left(j\omega L_2 + \frac{1}{j\omega C_2} + R_T\right) I_2 \tag{1.4}$$

Here, $R_T = R_2 + R_l$ is the total resistance of the secondary circuit. From Equation 1.3 and Equation 1.4, current in the second circuit can be rearranged as follows:

$$I_2 = \frac{j\omega M}{\left(R_1 + \frac{1}{j\omega C_1} + j\omega L_1\right) \times \left(R_T + \frac{1}{j\omega C_2} + j\omega L_2\right) + \omega^2 M^2} V_1 \quad (1.5)$$

From Equation 1.5, the transfer function between the circuits can be defined in view of the supply voltage of the primary circuit and the induced output current of the secondary circuit. When the two circuits are in a perfect resonant condition, the tuning of the capacitances in primary and secondary circuits should meet the conditions stipulated in Equation 1.6 and Equation 1.7, as shown in Figure 1.1(b), at the resonance frequency of $\omega_0$.

$$\frac{1}{j\omega_0 C_1} + j\omega_0 L_1 = 0 \quad (1.6)$$

$$\frac{1}{j\omega_0 C_2} + j\omega_0 L_2 = 0 \quad (1.7)$$

Using these two equations, we can properly tune the circuits for the maximum magnetic coupling. From Figure 1.1(b), we can write the following relationship shown in Equation 1.8 and Equation 1.9 at the resonant operating frequency $\omega$.

$$V_1 = R_1 I_1 - j\omega_0 M I_2 \quad (1.8)$$

$$V_2 = j\omega_0 M I_1 - R_T I_2 \quad (1.9)$$

By defining the complex impedances of each circuit as $Z_1$ and $Z_2$, these can be written as Equation 1.10 and Equation 1.11.

$$Z_1 = R_1 + \frac{1}{j\omega_0 C_1} + j\omega_0 L_1 \quad (1.10)$$

$$Z_2 = R_T + \frac{1}{j\omega_0 C_2} + j\omega_0 L_2 \quad (1.11)$$

For further analysis, let us represent the series combination of $C_1$ and $L_1$ through the equivalent impedance $Z_1$, and the series combination of $C_2$ and $L_2$ by $Z_2$, as $M$. For additional simplification of the analytical process, we make the assumption that both primary and secondary impedances are identical, that is $C_1 = C_2 = C$ and $L_1 = L_2 = L$. Thus, as shown in Equation 1.12,

$$Z_1 = Z_2 = j\left(\omega L - \frac{1}{\omega C}\right) \quad (1.12)$$

As a result, both resonators applied to primary and secondary coils have the same natural resonant angular frequency, as given by Equation 1.13.

$$\omega_0 = \frac{1}{\sqrt{LC}} \tag{1.13}$$

The application of Kirchhoff's voltage law around the primary and secondary loops leads to the equation describing the charging output voltage $V_2$, which can be maximized at two respective resonant angular frequencies, the magnetic resonant frequency $\omega_m$ and electric resonant frequency $\omega_e$, given in Equation 1.14 and Equation 1.15.

$$\omega_m = \frac{1}{\sqrt{(M+L)C}} = \frac{\omega_0}{\sqrt{(1+k)}} \tag{1.14}$$

$$\omega_e = \frac{1}{\sqrt{(L-M)C}} = \frac{\omega_0}{\sqrt{(1-k)}} \tag{1.15}$$

The power delivered to the secondary circuit $P_2$ and, hence, the transfer efficiency $\eta$ at the resonance frequency $\omega_0$ can be obtained by Equation 1.16 and Equation 1.17.

$$P_2 = \text{Re}\{V_2 \cdot I_2^*\} = \frac{\omega_0^2 M^2 R_T \cdot I_2}{Z_2 \cdot Z_2^*} \tag{1.16}$$

$$\eta = \frac{P_2}{P_1} = \frac{\text{Re}\{V_2 \cdot I_2^*\}}{\text{Re}\{V_1 \cdot I_1^*\}} = \frac{\omega_0^2 M^2 R_T}{\text{Re}\{Z_1 \cdot Z_2 \cdot Z_2^* + \omega_0^2 M^2 \cdot Z_2^*\}} \tag{1.17}$$

where "*" represents the complex conjugate. By simplifying Equation 1.17, transfer efficiency can be written as shown in Equation 1.18.

$$\eta = \frac{\omega_0^2 M^2}{R_1 \cdot R_T + \omega_0^2 M^2} = \frac{1}{1 + \frac{R_1 R_T}{\omega_0^2 M^2}} \tag{1.18}$$

The Q-factor is a quantitative representation of the sharpness of the circuit resonance. In power systems, the Q-factor is determined from a passive element, such as $R$, $L$, and $C$. The Q-factor of an inductance $Q_L$ and the Q-factor of capacitance $Q_C$ can be written as shown in Equation 1.19 and Equation 1.20.

$$Q_L = \frac{X_L}{R_L} = \frac{\omega_0 L}{R_L} \tag{1.19}$$

$$Q_C = \frac{X_C}{R_C} = \frac{1}{\omega_0 C R_C} \tag{1.20}$$

$Q_1$ and $Q_2$ indicate the Q-factors of the primary and secondary circuits, and these are incorporated into $Q_L$ and $Q_C$, respectively. Both Q-factors can be defined as shown in Equation 1.21 [1-10].

$$Q_{1,2} = \frac{Q_{L1,2} Q_{C1,2}}{\left(Q_{L1,2} + Q_{C1,2}\right)} \tag{1.21}$$

With $Q_{1,2}$, the power transfer efficiency (PTE) can be estimated by Equation 1.22 based on resonance sharpness.

$$\eta = \frac{k^2 (Q_1 \cdot Q_2) M^2}{1 + k^2 (Q_1 \cdot Q_2)} \tag{1.22}$$

Thus, to maximize the transmission efficiency as described in Equation 1.18, both resonant circuits should follow this basic principle [1-10].

$$\eta = \frac{R_1 R_T}{\omega_0^2 M^2} << 1 \tag{1.23}$$

From Equation 1.23, it should be noted that higher operating frequency and stronger magnetic coupling in both resonant circuits will bring higher power transfer efficiency. From Equation 1.2, it is understood that higher $M$ requires higher $k$, where one of the significant factors would be the magnetic field interaction between the two resonant coils.

### 1.1.2 General Application Field

The applications of WPT are gradually increasing throughout industries including consumer electronics, transport, and home appliances. Despite the potential increase in convenience, the breadth of applications of WPT is limited by the applicability and practicality of current technologies [1-7]. In this subsection, application types such as electronics, transport, energy harvesting, and medical implant are introduced (see Figure 1.2).

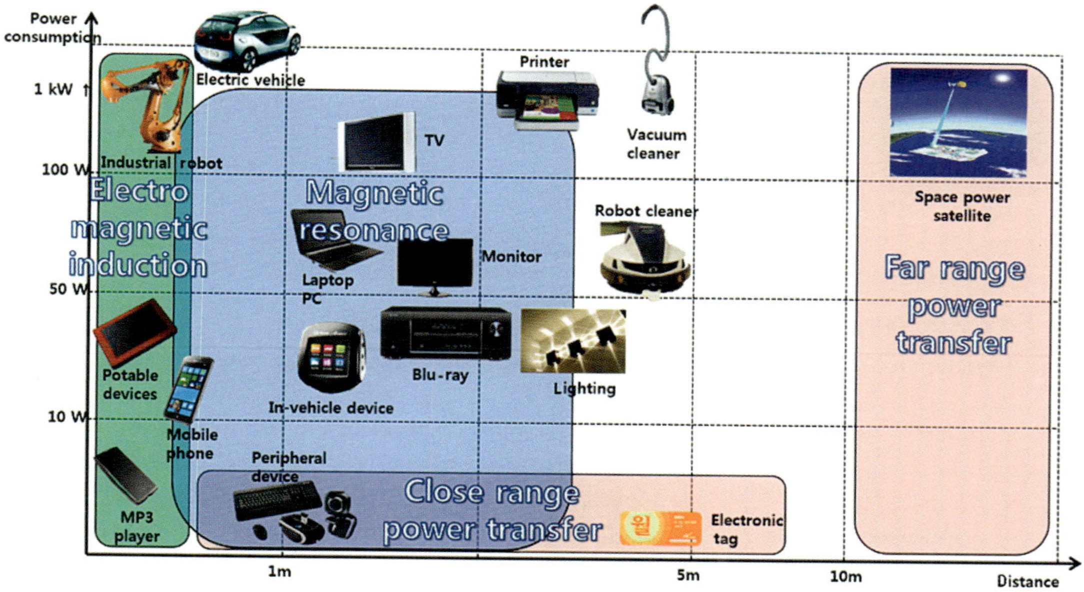


**Figure 1.2** Applications of WPT technology.

### *1.1.2.1 Consumer Electronics*

Portable devices such as smart phones, cameras, and laptops have limited battery capacity, so they must be recharged frequently. Most current portable devices can be charged by using a charger. This means that users need to carry a charger to be able to recharge portable devices. In addition, recharging portable devices could be inconvenient because the charger requires a connection with a power source like a plug, and the cable attached to any charger has limited length. The application of WPT technology can offer more convenient portable device recharging than conventional chargers that are connected by cable because WPT does not require a conductive connection.

For better understanding, this aspect of wireless charging can be equated with the development of information technologies. In the past, to access internet, LAN cables were used, but now we can access internet without cable connection by using wireless access such as WiFi and Long-Term Evolution (LTE). In other words, future wireless connections can provide not only data transmission but also power transmission without conductive connection.

Wireless charging areas are not limited to indoor applications. In the case of vehicles, with the development of vehicle technologies, linkage services with portable devices are being provided. Data connections between vehicles and portable devices connected by wireless interfaces such as Bluetooth exist, but charging of portable devices is usually done with a cable connection. If wireless charging is applied, data communication and power transmission between vehicle and portable devices can be operated simultaneously without any cables. These wireless charging environments provide more convenient portable device recharging, and contribute to portable devices with reduced battery capacities because they can be charged automatically in various places.

Moreover, the supply of electrical power without cable connections can also be applied to home appliances. Current home appliances such as televisions, clocks, refrigerators, and vacuum cleaners are supplied with electrical power through conductive cables. If wireless power transfer technologies are applied for supplying power to home appliances, complex conductive cable connections can be reduced, making home environments more safe and neat because of elimination of multiple plugs and cables.

#### 1.1.2.2 Industrial Automation and Heavy Equipment

One drawback of supplying electrical power through conductive materials is that it could be hazardous in wet environments. Most devices used in wet environments are designed as waterproof structures, but the connection point for supplying electrical power is hard to design as a waterproof structure. This means that short circuits caused by water can occur at electrical connection points and be detrimental to electrical safety. If a device requires a fully waterproof design, wireless power supply techniques can be applied to achieve waterproofing.

Typically, subsea instruments that are used to detect and extract natural gas and oil require waterproof designs because their operation environment is underwater. If wireless power supply techniques are applied to subsea instruments, electrical power can be transmitted safely because there would be no requirement for a conductive connection.

In addition, industrial automation designs can be enhanced by applying WPT technologies. Industrial automation systems are composed of a number of subsystems, which require an electrical power supply. This means that the industrial automation systems have complex electrical connections organized by conductive connection. Applying wireless power transfer to industrial automation can eliminate complex electrical connections so industrial automation systems can be safer than connections through cable.

The supply of power to heavy equipment used in mining, forestry, and remote areas is difficult because the power source is far away. This limitation would lead to delay of work and consequently, cost could be incurred. If WPT is applied for heavy equipment, electrical power can be supplied regardless of the power source so that equipment utilization would be more convenient than it presently is.

#### 1.1.2.3 Transports

Wireless charging methods for vehicles propelled by electrical energy can be classified into two types, as stationary and dynamic charge. Stationary charge is a type of electrical charge that supplies electrical energy only when the vehicles do not move. Plug type charging, as described in the SAE J1772 standard, should be plugged and unplugged by vehicle operators, and this type of charging is also hard to design as an automatic system. This means that vehicle operators can forget to charge their transports so that they cannot be used because of having no power.

Plug-in fast charging is dangerous for users because of its high voltages and currents. In addition, charging vehicles by using plug interfaces outdoors is also dangerous because rain and snow could short circuit the charger. These inconvenient and dangerous operations could be resolved by applying WPT techniques. WPT methods do not require a conductive connection, and so charging can be operated automatically by a controller to increase user convenience. In the case of fast charging, no user operation such as plugging and unplugging would be required, which would enhance electrical safety. Moreover, charging vehicles outdoors could be safer because WPT is not influenced by inclement weather such as rain or snow.

Dynamic charging is a type of charge that supplies electrical power while moving. An example of dynamic charging with conductive connection would be a pantograph charging system. It is a conductive connection usually installed on the roof of a vehicle. Pantograph charging has safety and applicability limitations for various types of transport. If a vehicle has a flat structure, the power line should be installed at an area near the ground. This can be dangerous for users because of high voltages on power lines. If power lines are installed far from the ground, the system could only be used by vehicles that have the same height, reducing its usefulness.

On the other hand, WPT based dynamic charging can provide safety because there is no way for users to come in contact with power lines. In addition, WPT power lines are installed at ground level so that they can be used by various types of transport regardless of their height. Dynamic charging can supply electrical power while driving so that vehicles do not require high battery capacities. Reducing the battery capacity installed in a transport can eliminate unnecessary power spending, and contributes to enhancing efficient energy utilization.

#### 1.1.2.4 Medical Implants

Human organs can lose functionality over time, decreased from senescence and/or functional loss from accidents. In order to resolve functional insufficiency in organs, medical implants are being developed and applied. A medical implant is a kind of prosthesis for supporting functional performance of an organ. Many implantable artificial organs use electrical energy for continuous operation. Various medical implants have been introduced such as pacemakers, cochlear implants, functional electric stimulators, and artificial hearts. The biomedical applications include diagnosis-purpose or life-aids, such as medical capsule endoscopy, physiological monitoring devices, and ventricular assist devices.

These medical implants require electrical energy for operation, and electrical energy is supplied from a battery installed in the medical implant. Batteries must be charged, because they cannot produce electrical energy themselves. The conventional medical implant recharge method is to take the medical implant out of the body through a surgical operation. However, this is dangerous because surgical operations always have risk. These limitations in medical implant charging can be resolved by applying WPT.

If WPT is applied to medical implants, electrical power can be transferred from an external power source to the medical implant through human skin. Currently, some medical implant charging technologies based on inductive power transfer have been introduced, but their power transfer efficiency is too low for convenient use. This low power transfer efficiency is caused by the complexity of the structure of a human body because it is composed of multiple organic mediums including skin, bone, muscle, blood, etc. For this reason, WPT techniques for medical implants require additional research before they can be properly applied to human bodies, with enhanced power transfer efficiency. In addition, to increase reliability and reduce health risks, a comparison to other WPT applications should be researched because medical implants are embedded into human bodies.

### 1.1.3 Technology Development Potential

Wireless power supplies are intended as components for other electronic products made by original equipment manufacturers, including batteries, chipsets, cellphones, cameras, digital media players, GPS devices, laptops, TVs, industrial wireless sensors, and electric vehicles. The total market for OEMs of wireless power transfer technology are forecasted [1-7]. In this forecast, WPT market size in the United States in 2010 is estimated around $70 million.

More specifically, two companies, Powermat and Pure Energy Solutions, have been selling consumer-oriented products since October 2009. These products are in the form of charging pads or mats that wirelessly recharge batteries in cellphones set on or near the pad. From these commercializations, low-power conductive and magnetic inductive consumer chargers are recognized as commercially available products by regulators. Wireless power supplies as components for embedding in product-specific designs entered the market in 2012, and wireless power consumer electronics are expected to continue to increase their share of the market. In addition, magnetic resonance WPT technologies that deliver power wirelessly over longer distances compared to pure inductive WPT were introduced. The products applying the magnetic resonance principle would contribute to diffuse WPT technologies at a faster pace. Consequently, the scale of market is expected to reach about $617 million by 2015. However, challenges related to business, technical issues, regulations, and standards still exist.

As for business challenges, many fundamental issues are related to OEMs, such as setting price points and estimating how many units of a wirelessly powered product they can sell. The sale of a few hundred thousand units isn't sufficiently attractive for OEMs to make the necessary investment. Volumes upwards of several hundred thousand units are necessary for production costs to come down and for consumer awareness of a technology to become sufficiently high for widespread adoption of a technology to take place.

Technical and regulatory questions also exist. Cellphones and handheld mobile devices introduce their own set of issues to be addressed. To deliver power far away from the original power supply, wireless power transfer has to be operated by high frequency. Miniaturization of the receiving coil is also necessary because of the limited mobile device form factor. With this constraint, a high operating frequency is applied to these systems. This is problematic from an electrical engineering standpoint because smaller components become very costly, as well as from a regulatory standpoint because higher frequencies bring concerns about the health risks of EM energy.

As for standardization, OEMs want to know that the wireless power solution being offered to them is something the consumer already wants. In competitive technology markets, standards are generally thought to be good for the industry and good for the consumer, and the same expectation is held for wireless power technology. When a component can be offered as an interoperable, standard-compliant solution, it represents a roadmap to lower costs for OEMs and keeps the consumer from having to pay a lower price for getting that technology. When it comes to commercializing wirelessly charged end-products, OEMs want to ensure that different cellphones and laptops can share the same charging equipment. Wireless power interoperability across rechargeable electronic devices is expected before manufacturers start full production of wirelessly powered product lines.

## 1.2 Wireless Power Transfer in Transportation

Electrification in transportation technology has been emphasized strongly for the last several decades and will continue to grow due to several factors driving the change in power sources, such as tougher regulations triggered by environmental concerns and an urge to reduce the crude oil dependence caused by energy security concerns. In automotive applications, the internal combustion engine (ICE), hybrid, compressed natural gas (CNG), and fuel-cell-powered vehicles will all be core power sources in current and future ground vehicle technology, an area that requires rapid enhancement of fuel efficiency by 2020 [1-11]. Despite the current slow pace of market penetration, electric vehicle (EV) technology is one of the strongest candidates in the industry and research sectors. While rapid progress is currently being made in the global electric vehicle market, substantial barriers to massive EV adoption still exist. Those can be summarized as batteries, charging infrastructure, electric power interface, and customer acceptance [1-12]. With the limited lithium-ion battery technology, the power density and specific power capacity have grown significantly in recent years. However, batteries are still unable to compete with the tremendous energy density of petroleum fuels [1-13].

Carrying the energy storage system (ESS) within an electric vehicle during the entire operation distance has been a significant roadblock to EV development, mainly because of the heavy weight and bulk of current battery systems. Overhead wire systems for traditional public tram can be a remedy, but they are limited to urban public transportation and obstruct scenic views on city streets. For these roadblocks, EVs delivering the power wirelessly to operating vehicles can be a competitive design solution for future electrified roads and vehicles [1-14]. Thus, wireless charging of EVs, whether stationary or dynamic on-road charging, can be a technological innovation to achieve widespread EV introduction. However, this technology should meet the following requirements: transfer power capacity and efficiency and an appropriate level of convenience, safety, and business competitiveness.

The development history of wireless power transfer technology can be summarized with several novel innovations. In the 1890s, in the Wardenclyffe tower experiment, N. Tesla boldly proposed the idea of transmitting electrical power wirelessly through the air [1-15], [1-16]. With increased power and energy-efficient consumer electronics and the use small-capacity mobile equipment, WPT technology has drawn significant attention in technology groups and related industries, and the market forecast predicts about $2 billion USD in wireless-power related revenues by 2020 in the areas of consumer electronics, automotive/transportation, industrial automation and heavy equipment, energy, sensors and transducers, medical and healthcare, and other applications [1-7].

In the WPT field, three main types of WPT are commercially viable at this time: electromagnetic inductive coupling, resonant magnetic coupling, and microwave-based wireless power transfer. NASA's power beaming by laser concept and space-based solar power are also noticeable in the WPT field, but both are either at the conceptual level or too immature for commercial application [1-7], [1-17], [1-18].

In the inductive coupling area, in addition to many application technologies developed in the consumer electronics area, the California PATH program and a group of researchers in Oakland University, New Zealand, have made a noteworthy contribution related to surface transportation and electric vehicle charging [1-19], [1-20], [1-21], [1-22]. The wireless power transfer experiment from the PATH program can be summarized with a transferred power capacity of 60 kW, operating frequencies of inductive power transfer of 400 Hz and 8500 Hz, and maximum measured power transfer efficiency from a commercial electricity source to the output of a paired receiver system of 60% with a 2–3 inch air gap [1-21], [1-22]. In addition, the PATH program includes various parametric simulation studies on the efficiency, economic analysis, and issues related to the exposure of humans to electromagnetic fields. Although the PATH program showed technological possibility of wireless power transfer in limited condition, the program also showed limitations of commercialization due to the low power transfer efficiency and poor competitiveness in business calculations.

Boys and Covic introduced a contact-less energy transfer system with a static battery charging system called inductively coupled power transfer (ICPT) [1-23], [1-24]. Soljacic et al. [1-8] successfully mathematically formulated the resonant coupled power transfer between two resonant coils, demonstrating the transfer of 60 W at a distance of 1.8 m using 10 MHz ac electricity.

In 2009, in Korea, the KAIST On-Line Electric Vehicle (OLEV) project group showcased a series of buses and a tram demonstrating the dynamic wireless charging of EVs as a commercially viable approach. The dynamic wireless charging of the OLEV system was possible by introducing resonant-based power transfer between primary and secondary coils. With the relatively larger distance between two coils, resonant coupling enabled dynamic charging in terms of power transmission efficiency, transferred power capacity, and economic considerations.

### 1.2.1 WPT in Automobile Applications

Two types of WPT can be applied to automotive charging: stationary charging, which is a technology that can provide wireless charging to a vehicle while stopped, and dynamic charging, which can provide wireless charging whether the vehicle is driving or at a designated stop. Stationary wireless charging is generally applied in parking lots. The operating mechanism of stationary wireless charging is shown in Figure 1.3 [1-25]. A further detailed description will be given in chapter 3.

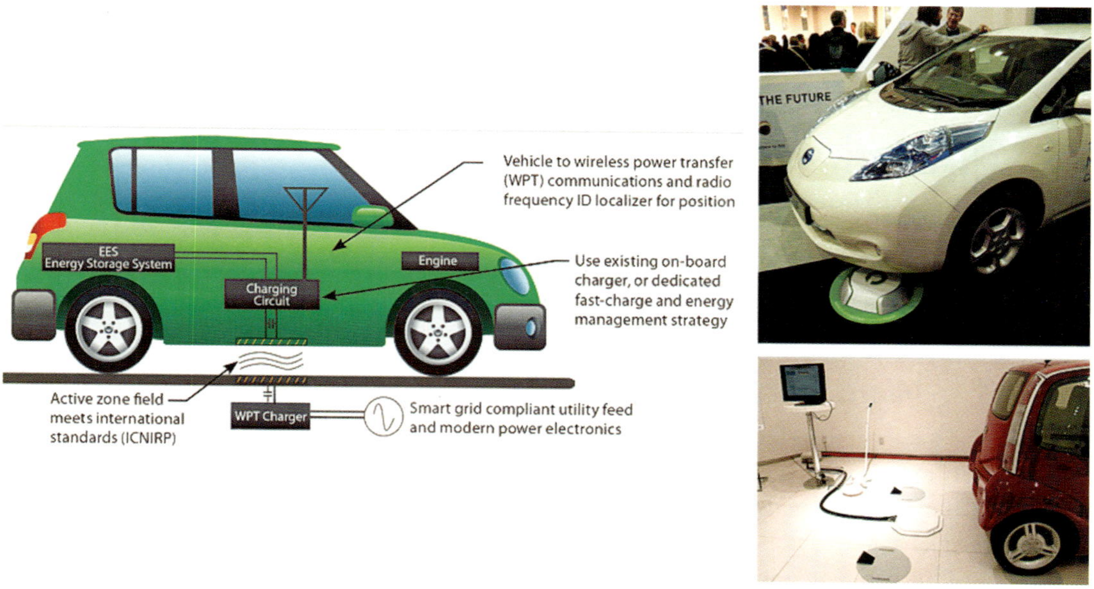


**Figure 1.3** Concept of stationary wireless charging [1-25].

Dynamic wireless charging technology, shown in Figure 1.4, is similar to stationary wireless charging. However, some differences do exist. First, the power supply module is installed under the road, and a number of power transmitting segments for generating an electromagnetic field have to be installed. A series connection is used because a vehicle would pass over each power segment for only a short amount of time. In addition, sequential communication between the power supply system and vehicle is required. Power segments are organized in sections consisting of clusters of power segments. Sections of road to be powered are selected based on evaluation of their economic feasibility.

Both wireless charging technologies, stationary charging and dynamic charging, have to satisfy minimum efficiency requirements in order to supply sufficient power. To achieve high efficiency, electrical and mechanical considerations should be considered and applied to the WPT system.

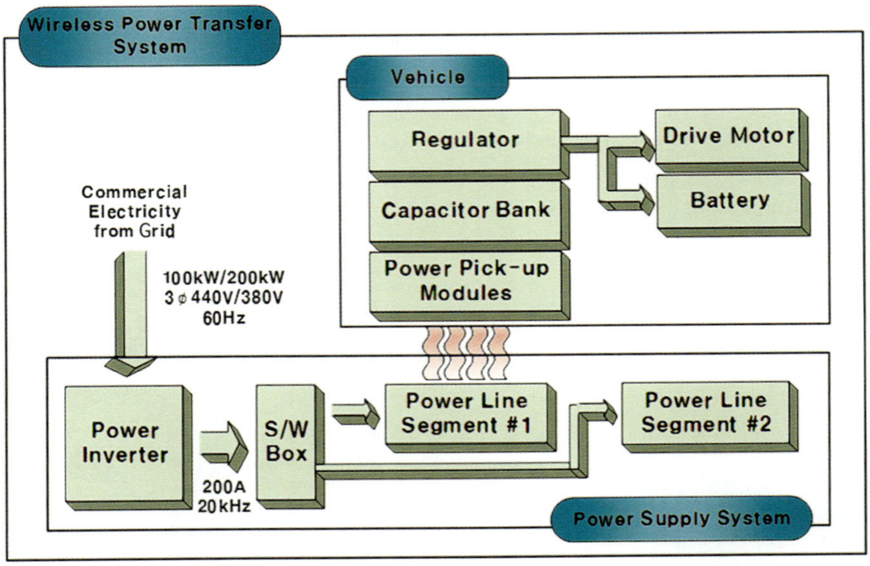


**Figure 1.4** Concept of dynamic wireless charging.

## 1.3 Micro Mobility and Wireless Power Transfer

Traffic congestion and parking lot shortages in megacities due to increasing population density are also serious transportation problems. Current global trends in population distribution are migrating toward large cities [1-26]. According to this trend, about 60% of the global population will reside in cities by 2017. In the case of Europe and Japan, 70% of the population is expected to reside in cities. The continuously increasing number of vehicles and population density in cities are heading toward serious traffic congestion and parking lot shortages. The above-described global population trends are hard to accommodate by existing transportation systems. New mobility and transportation systems are required to deal with these transportation-related challenges effectively [1-27]. Among these challenges, an aging society is one of the key elements that needs to be dealt with in transportation system planning. According to the statistical data of the Organization for Economic Cooperation and Development (OECD), in some OECD countries, the population ratio of those above 65 years old is expected to lead to an aging society by 2020 [1-28]. Several nations already have significantly aged populations. A few examples are Japan (26.2%), Germany (21.6%), France (20.1%), and Sweden (23.1%). This means that issues related to people with reduced mobility (PRM), such as elderly, disabled people, and pregnant women, are expected to increase significantly in the mid- and long-term. A new type of mobility should be created to address these concerns.

Personal mobility vehicles (PMVs) or micro electric vehicles (MEVs) that use electricity as their main source of power are drawing attention globally as possible solutions to some of these problems. PMVs are expected to provide various benefits such as pollution reduction, opportunities for a new business model, and mobility for the disabled and elderly. However, the global regulation of PMVs is not clearly defined yet. A variety of PMVs exists ranging from one-person mobility vehicles, such as the Segway and Solowheel, to two-seater micro vehicles, and PMVs are usually operated by electric propulsion. Among these PMVs, four-wheel micro mobility vehicles such as Twizy, which was introduced by Renault, are being developed, so the market size is gradually increasing. PMVs can provide additional benefits to the rapidly growing numbers of elderly, car-sharing applications, convenient transfer within stations or inter-modal applications, local city tours, and other short-distance driving areas of urban transport. Figure 1.5 illustrates PMV applications for public use.

**Figure 1.5** Various PMV applications for public use.

A demonstration program was funded by Oak Ridge National Laboratory (ORNL) to showcase the benefits of wirelessly charging a neighborhood electric vehicle, a Global Electric Motors (GEM) 4-passenger, low-speed EV. In-motion wireless charging consists of a sequence of roadway embedded coils, an energized track in which individual coils or pairs are sequentially energized by a track-side high-frequency power inverter in synchronization with vehicle position. The ORNL in-motion charging demonstrator is shown schematically in Figure 1.6 and has the following key functions:

- Primary coils connected in pairs and in-phase (i.e., fountain field), singly tuned to 22 kHz.

- Coil sequencing controlled by vehicle passage using track-side photocell interruption.
- Power level controlled by high-frequency (HF) inverter rail voltage.
- Single secondary, or capture coil, centrally mounted to demo vehicle chassis.
- Maxwell Technologies ultra-capacitor pack in passive parallel with demo vehicle battery pack.
- Electro Standards Laboratories (ESL) lithium-capacitor rack in active parallel with grid-side power inverter. This apparatus was designed for both passive and active parallel connection.
- Zlinx radio communications between vehicle and grid-side controller.

Figure 1.6 shows the ORNL wireless charging laboratory set up with an in-motion test track consisting of a six-coil electrified portion, the GEM EV, and laboratory instrumentation; an ESL equipment rack containing lithium-capacitors (LiC) to smooth the grid power is in the foreground. During this test, the Maxwell Technologies carbon ultra-capacitor was in passive parallel with the vehicle lead-acid battery, and the ESL LiC was in active parallel with the grid-side power supply. The ESL equipment rack contained a dedicated power supply along with a dc-dc converter to interface the LiC in active parallel, all designed to mate with the ORNL track-side electronics.

**Figure 1.6** Dynamic testing in-motion wireless charging at ORNL using electrochemical capacitor smoothing (Mr. Cliff White shown monitoring grid-side equipment and Mr. Steven Campbell driving the test vehicle).

The light poles shown in Figure 1.6 on the grid-side HF current sequencing electronics box and on the GEM vehicle are connected for visual indication of grid power smoothing and smoothing of the vehicle battery pack charging current. Both visual indicators showed dramatic reduction in power pulsation when power capacitors were used, as described in detail in literature [1-29].

Figure 1.7 shows ORNL's concept for dynamic WPT that would be installed in special charging lanes on highways so that specially equipped vehicles can charge as they drive. The key attributes of this concept are that WPT modules are controllable, modular, scalable, and replaceable, and the energized track sections are powered by highway dc-distribution and are robust to failure of multiple points. Coil sequencing is synchronous with the vehicle under charge regardless of speed or congestion. WPT roadside units (RSU) supply the dc distribution from grid connections at various intervals along the highway corridor.

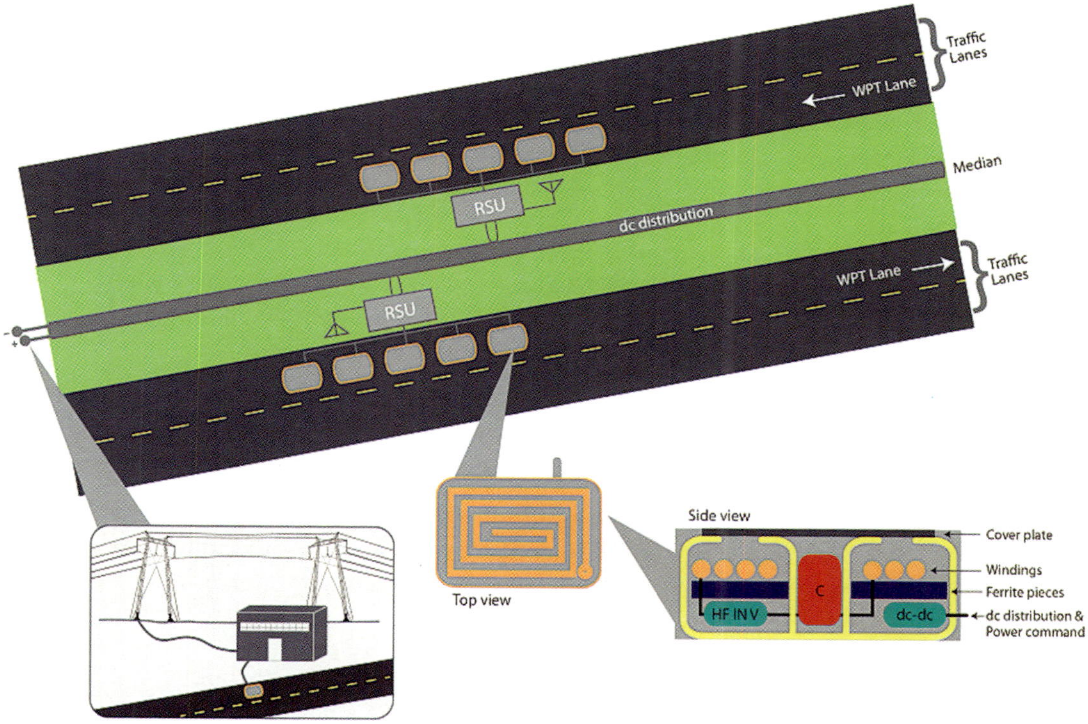


**Figure 1.7** Modularized dynamic WPT for future in-highway installation.

Wireless charging of PEVs is a concept that can be implemented in several ways, with the modularized concept shown in Figure 1.7 being just one and the OLEV being another viable method. There will be many challenges in integrating in-motion or dynamic WPT into public highways, but these will be solved through engineering and the application of materials necessary for long-term survival in such a demanding application.

## 1.4 Structure of the Book

This book is intended to provide an overview of the current technology status of wireless power transfer technology and applications to the future transportation field, including other consumer electronics. The structure of this book is as follows. In chapter 1, the basic principle of inductive power transfer technology is reviewed, along with different types of power transfer technologies and applications. For future sustainable transportation, the importance and feasibility analysis of WPT application to electric vehicles are discussed as well. In chapter 2, discussions on future vehicle technology, emphasizing the automotive propulsion change toward electric vehicles, are discussed. Combining the progress of research and development in the intelligent transportation system, with the emphasis on collaborative and connected driving, the power source change in traction and WPT technology can be integrated with the building block of the future road transportation system.

In chapter 3, the charging technology developments on conductive and wireless charging for electric vehicles are discussed. Implementing innovative WPT technology to the practical commercial products and roll-out strategies are also discussed in this chapter. Among the WPT application cases, the unique contribution by KAIST on the dynamic wireless charging of the electric bus are described in view of its basic working principle, performance specification, and application scenario, etc., in chapter 4. The OLEV system characteristics are described, including commercialization phase considerations.

In chapter 5, technical issues and challenges for wireless power transfer technology are described, including perspectives on alignment between the supply module and receiver module as well as operation, control, and handling in emergency circumstances.

In chapter 6, discussions on the market strategies and technical standards on wireless EV charging are introduced, considering the current and future market potential, and the design and implementation of charging infrastructures; practical applications are also described. With the growing emphasis on IT technologies in the automotive industry, wireless charging technologies for mobile or IT devices are discussed in chapter 7.

Chapter 8 focuses on WPT application in railway, focused on metro, light rail, high-speed rail, and low-floor tram, respectively. Long-distance power transfer is described in chapter 9 with its history, theory, and recent applications. Last, disruptive technologies for applying WPT, focusing on aviation, marine, off-road, and commercial applications, are introduced in chapter 10.

## References

1-1. Marian P. Kazmierkowski and Artur J. Moradewicz, "Unplugged but connected: Review of Contactless Energy Transfer systems," IEEE *Industrial Electronics Magazine*, Dec. 2012.

1-2. R. M. Dickinson, "Wireless power transmission technology state of the art: The first Bill Brown lecture," *Acta Astronautica*, vol. 53, nos. 4–10, pp. 561–570, 2003.

1-3. D. E. Raible, D. Dinca, and T. H. Nayfeh, "Optical frequency optimization of a high intensity laser power beaming system utilizing VMJ photovoltaic cells," in Proc. Int. Space Opt. Syst. Appl. Conf., May 2011, pp. 232–238.

1-4. A. Sahai and D. Graham, "Optical wireless power transmission at long wavelengths," in Proc. Int. Space Opt. Syst. Appl. Conf., May 2011, pp. 164–170.

1-5. C. Liu and A. P. Hu, "Steady state analysis of a capacitively coupled contactless power transfer system," in Proc. IEEE Int. Conf. Energy Convers. Congr. Expo., San Jose, CA, 2009, pp. 3233–3238.

1-6. C. Liu, A. P. Hu, and N.-K. C. Nair, "Modelling and analysis of a capacitively coupled contactless power transfer system," *IET Power Electronics*, vol. 4, no. 7, pp. 808–8015, 2011.

1-7. *Wireless Power: Current Status and Future Directions*, Fuji-Keizai USA Inc., Oct. 2010.

1-8. A. Kurs, A. Karalis, R. Moffatt, J. D. Joannopoulos, P. Fisher, and M. Soljacic, "Wireless power transfer via strongly coupled magnetic resonance," *Science Express* vol. 317, no. 5834, pp. 83–86, 2007.

1-9. I. Suh and J. Kim, "Electric vehicle on-road dynamic charging system with wireless power transfer technology," 2013 IEEE IEMDC, pp. 247–253, May 2013.

1-10. J. H. Kim, "Resonant magnetic field design for high efficiency and low EMF wireless power transfer," 2010 International Forum on Electric Vehicle (2010 IFEV), Daejeon, Korea, 2010.

1-11. Georg Erdmann, Jürgen Kluge, Philipp Radtke, and H. Wallentowitz, *DRIVE-The Future of Automotive Power*, McKinsey & Company, Inc., 2006.

1-12. *Electrification Roadmap*, Electrification Coalition, 2009.

1-13. I. S. Suh, "Application of SMFIR Technology to Future Urban Transportation," in 21st CIRP Design Conference, Daejeon, Korea, 2011.

1-14. N. P. Suh, D. H. Cho, and C. T. Rim, "Design of On-Line Electric Vehicle," in Plenary Presentation at 20th CIRP Design Conference, Nantes, France, 2010.

1-15. N. Tesla, "Art of transmitting electrical energy through natural medium," US Patent 1119732, 18 Apr. 1905.

1-16. N. Tesla, "Apparatus for transmitting electrical energy," US Patent 1119732, 1 Dec. 1914.

1-17. "Wireless Power Transfer Technology Workshop Proceedings," Korea Institute of Electrical Engineers, 2011.

1-18. Y. Hori and Y. Yokoi, *Wireless Power Transfer and Infrastructure Construction for Electric Vehicles*, Japan: Shemusi Press, 2011.

1-19. S. E. Shladover, "Highway Electrification and Automation," UC Berkeley PATH, California, 1992.

1-20. E. H. Lechner, S. E. Shladover, K. Lashkari, and D. M. Empey, "Inductive energization system and method for vehicles," US Patent 5207304, 4 May 1993.

1-21. "Roadway Powered Electric Vehicle Project Track Construction and Testing Program Phase 3D," California PATH Research Report, pp. UCB-ITS-PRR-94-07, Mar. 1994.

1-22. "Roadway Powered Electric Vehicle Project Parametric Studies: Phase 3D Final Project," California PATH Research Report, pp. UCB-ITS-PRR-96-28, Oct. 1996.

1-23. C. W. Wang, O. H. Stielau, and G. Covic, "Design considerations for a contactless electric vehicle battery charger," IEEE Transactions on Industrial Electronics, vol. 52, no. 5, pp. 1308–1314, Oct. 2005.

1-24. G. H. Covic, G. Elliot, O. H. Stilau, R. M. Green, and J. T. Boys, "The design of a contact-less energy transfer system for a people mover system," in International Conference on Power System Technology Proceedings, Auckland, 2000.

1-25. Oak Ridge National Laboratory (ORNL), "Wireless Charging System for Electric Vehicles," Sep. 2011.

1-26. T. V. Riper, The 10 Biggest Cities Of 2025, http://www.forbes.com/2008/03/19/cities-population-pollution-innovation08-cx_tvr_0319futurecities.html, 19 Mar. 2008.

1-27. United Nations, *The State of Chinese Cities, 2010*, Habitat for a better urban future, Foreign Languages Press, Beijing, 2010.

1-28. Organization for Economic Cooperation and Development, www.OECD.org, 2013.

1-29. J.M. Miller, O.C. Onar, C.P. White, S.L. Campbell, C. Coomer, L. Seiber, R. Sepe Jr., and A. Steyrl, "Dynamic Wireless Charging of Electric Vehicle Demonstrated at Oak Ridge National Laboratory: Benefit of Electrochemical Capacitor Smoothing," for publication in *IEEE Power Electronics* Magazine, first issue 2014.

# Chapter 2
# Green Transportation and Electric Vehicles

## 2.1 Future Automotive Power Drive Trend

In 2010, the International Energy Agency (IEA) performed the most comprehensive analysis of the world energy future, conducted by the IEA Committee on Energy Research and Technology (CERT). This study makes major recommendations for more sustainable energy future under the Accelerated Technology (ACT) Scenario and the Blue Map Scenario, compared with the Baseline Scenario (business-as-usual scenario). Under the Baseline Scenario, energy consumption will be doubled by 2050, and $CO_2$ emissions will continue to rise to 2.5 times the current levels. In the ACT scenario, energy consumption will be at about 77% of the Baseline Scenario, and energy-related $CO_2$ will be reduced to their 2005 levels by 2050. In the Blue Map Scenario, energy consumption will be at about 67% of the Baseline Scenario, and the $CO_2$ emissions will be 50% lower than the 2005 levels. This scenario explores the least-cost solutions to achieve this goal and to limit the risks of severe climate change. With three scenarios, IEA reported that hybrid electric vehicles (HEVs), including plug-in hybrid electric vehicles (PHEVs), would dominate the majority of the market by 2035 [2-1]. The share of pure electric vehicles (EVs), such as battery electric vehicles (BEVs) and fuel-cell electric vehicles (FCEVs), are expected to increase after 2035, and it is forecasted that the market share would be almost equal to that of 2050, as shown in Figure 2.1, under the Blue Map Scenario.

Statistical data show that PHEVs and EVs are beginning to increase from 2010 to penetrate the market, with EVs reaching sales of 2.5 million vehicles per year by 2020 and PHEVs reaching sales of nearly 5 million by 2020 (Figure 2.2, Table 2.1). By 2030, sales of EVs are projected to reach 9 million and PHEVs are projected to reach nearly 25 million. After 2040, sales of PHEVs are expected to begin declining as EVs (and fuel-cell vehicles) are predicted to achieve even greater levels of sales. The ultimate target is to achieve 50 million sales of both types of vehicles annually by 2050.

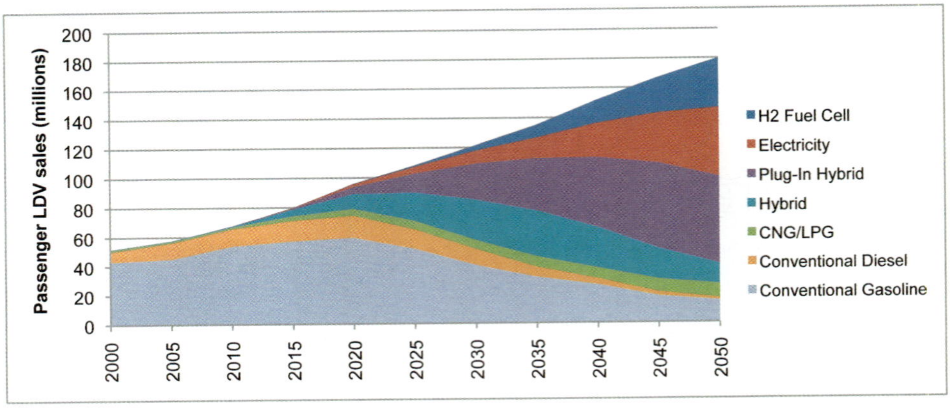


**Figure 2.1** Annual light-duty vehicle sales by technology type, Blue Map scenario (© 2011 OECD/IEA. Reprinted, with permission, from [2-1], IEA Publishing. License http://www.iea.org/t&c/).

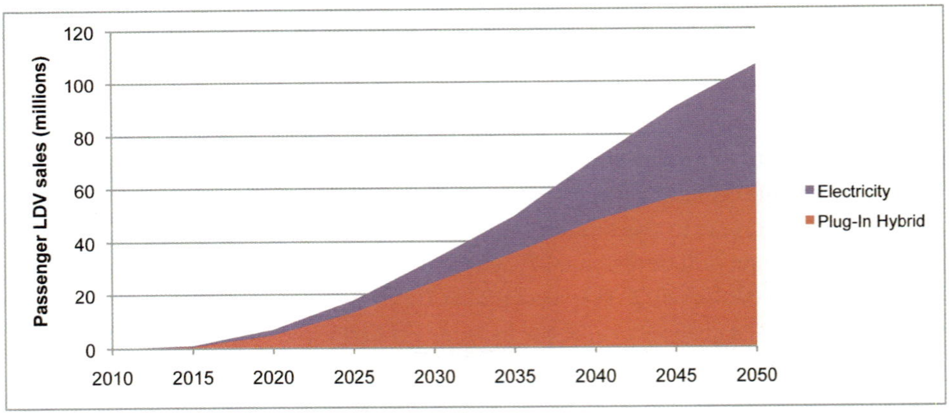


**Figure 2.2** Annual global EV and PHEV sales in the Blue Map scenario (© 2011 OECD/IEA. Reprinted, with permission, from [2-1], IEA Publishing. License http://www.iea.org/t&c/).

**Table 2.1 Global EV and PHEV sales in Blue Map (© 2011 OECD/IEA. Reprinted, with permission, from [2-1], IEA Publishing. License http://www.iea.org/t&c/)**

| | 2010 | 2015 | 2020 | 2025 | 2030 | 2035 | 2040 | 2045 | 2050 |
|---|---|---|---|---|---|---|---|---|---|
| *PHEV* | *0.0* | *0.7* | *4.9* | *13.1* | *24.6* | *35.6* | *47.7* | *56.3* | *59.7* |
| *EV* | *0.0* | *0.3* | *2.0* | *4.5* | *8.7* | *13.9* | *23.2* | *33.9* | *46.6* |
| *Total* | *0.0* | *1.1* | *6.9* | *17.7* | *33.3* | *49.5* | *70.9* | *90.2* | *106.4* |

Unit: million per year

It is important to note that for the near- to medium-term (2010 to 2020) data in the figures above, the IEA 2010 Blue Map scenario was revised in 2009 to account for both the economic crisis that began in 2008, which decreased projected car sales, as well as for PHEV/EV product plans announced since the Energy Technology Perspectives (ETP) was published, which suggest the possibility of higher levels of EV sales by 2020 [2-1]. This is an ambitious but plausible scenario that assumes strong policies and clear policy frameworks, including the provision of adequate infrastructure and incentives.

While it may be possible to reach $CO_2$ targets in other ways, if this target level of EVs and PHEVs relying on low-carbon electricity is not introduced, then other low $CO_2$-emitting solutions will be needed. Altering the Blue Map strategy in this way will likely result in an equally or even more difficult challenge.

### 2.1.1 Motivation

The growth rate of the global automotive market was recorded to be 24% in 2007, and this rate is expected to grow up to 56% by 2030 [2-1]. In other matters, global warming from $CO_2$ emissions has been a serious issue to consider within the transportation industries because studies show they have been responsible for about 25% of $CO_2$ emissions [2-2]. The development of an electric vehicle is a response to various problems such as air pollution, global warming, and shortages of energy resources [2-3]. To keep up with environmental goals, global regulations such as the Kyoto Protocol, Copenhagen Climate Change Conference, and Asia-Pacific Partnership are attempting to lower $CO_2$ emissions to preserve the environment. Accordingly, the need for environmentally friendly transport is continuing to increase with the growing concern over environmental issues, as summarized in Figure 2.3.

**Figure 2.3** Development motivation of environmentally friendly vehicles.

In addition, technologies that can replace the internal combustion engine (ICE) are needed because of limited natural resources such as fossil fuels. According to a report composed by the American Petroleum Institute, petroleum reserves are expected to be depleted between the years 2062 and 2094 [2-4]. For this reason, the problems described here must be resolved for future sustainability. Therefore, environmentally friendly transport technologies are being developed increasingly throughout the automotive industries.

### 2.1.2 Automotive Industry

The automotive industry has been facing numerous challenges, which can be summarized as a mega trend. The mega trend can be described with four major focuses, such as sustainability, globalization, changes in consumer behavior, and demographic development [2-5]. In the sustainability field, which is driven by the increasing environmental concerns and global climate changes, the reduction of global $CO_2$ emission savings targets are specified. As a result, the new energy sources, energy storage technologies, and their operation toward improving overall efficiency have been emphasized. In the globalization field, the automotive industry cannot be an exception to the increasing economic and political integration, increasing global technology networking, and sharing the related information via SNS world. The consumer behavior changes and its dynamics are another challenge to the future automotive industry. Due to the progress of global market development, the consumers' ethnicities, lifestyles, and preferences become diverse. Growing young population in developing countries, increasing aging population in developed countries, and rapid growing urbanization will be the noticeable trends in the demographic development.

### 2.1.3 Key Elements of Future Automotive Technology

The key elements of future automotive technology for market potential will be summarized in four categories: modularization, embedded electronic systems, low-cost high-tech, and electrification of the powertrain [2-6]. Modularization is a technology that separates vehicle parts and then recombines them in categorized form so that it is possible to make a complete car by assembling these parts, like LEGOs®. This could enable customization according to the customer's preferences, reducing the role of OEMs.

Vehicle-embedded electronic systems are the technologies related to functions for safety and convenience. The number of electronic parts equipped in the vehicles is continuously increasing to improve safety and provide convenience to the driver. The purpose of vehicle electrification can be summarized in three steps [2-7]. The first relates to active safety, which prevents accidents by alerting drivers to hazardous circumstances. If the driver does not react, the active safety system can control the vehicle independently so that an accident may be avoided. The second relates to connected vehicle technology, which provides convenience and safety to the driver by communicating with other vehicles or infrastructure, termed vehicle-to-vehicle (V2V) and vehicle-to-infrastructure (V2I), respectively. V2V and V2I use information communication technology (ICT) so that they can provide not only convenience and safety to the driver, but also accessibility to information from nomadic devices such as smart phones and smart pads. The third step is autonomous driving technology, which can drive a vehicle without the driver's control by using autonomous driving systems including sensors, algorithms, and controllers. The functions of autonomous technologies are to facilitate autonomous parking, vehicle platooning, and autonomous driving, among others.

The purpose of low-cost high-tech is to provide high technical functionality or performance systems at a low cost to customers. Current EVs and their parts are somewhat expensive because the technological development remains insufficient. In addition, most of the future vehicle technologies are applied to premium sedans because of their high technology cost, which will cause delays in vehicle technology diffusion. As a result, advanced technologies cannot reduce accidents or pollution even though the technology has been fully developed.

Powertrain electrification implies not only electric vehicles but also the energy generation. Current electrical energy is generated through turbine rotation by the combustion of natural resources or the generation of nuclear power. Table 2.2 shows the fractional source of world energy as of 2009 in the electricity generation and energy production [2-8]. This has raised a concern that current electric vehicles are not a perfectly ecological mode of transport. To resolve this limitation, environmentally friendly power generation technologies using renewable energy sources, such as photovoltaic, wind, hydro, and biomass, are being researched and developed. These technologies definitely have to be developed with EV technology for reducing environmental impact and natural resource consumption.

**Table 2.2 Fractional Source of World Energy as of 2009 (2011)**

| | Coal | Oil | Natural Gas | Nuclear | Hydro | Other | Total |
|---|---|---|---|---|---|---|---|
| *Electricity Generation* | *40.5%* | *5.1%* | *21.4%* | *13.4%* | *16.2%* | *3.3%* | *100%* |
| *Total Energy Production* | *28.1%* | *32.4%* | *20.6%* | *5.7%* | *2.3%* | *10.9%* | *100%* |

## 2.1.4 Classifications of the Environmentally Friendly Vehicles

Environmentally friendly vehicles (EFV) can be classified by the amount of electrical energy used for the propulsion of vehicles. All types of EFV use electric motors operated by electrical energy, but each EFV has a different dependency on electrical power for propulsion, as shown in Figure 2.4. Accordingly, various types of environmentally friendly vehicles exist, including BEV, HEV, PHEV, and FCEV, and those vehicles can reduce $CO_2$ emissions compared to ICE vehicles [2-9].

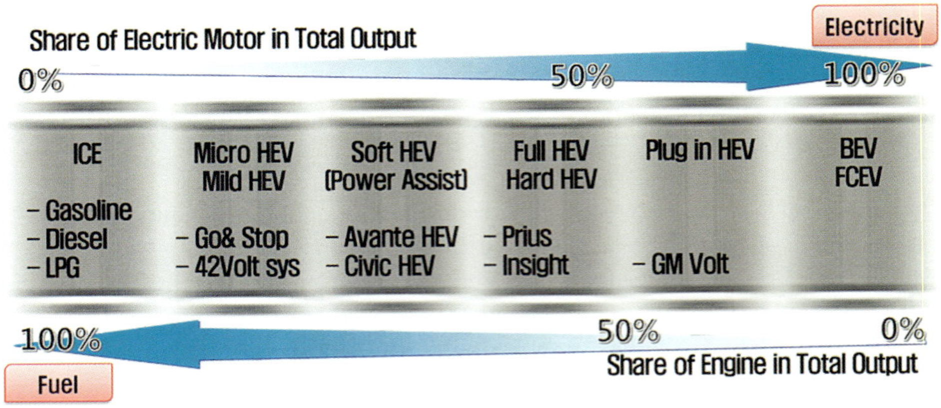


**Figure 2.4** Classifications of EFV (© 2010 IEEE. Reprinted from [2-9]).

A hybrid electric vehicle is one that combines a conventional internal combustion engine propulsion system with an electric propulsion system. Hybrid electric vehicles can be classified into four subcategories: mild HEV, in which the engine is shut down when the vehicle stops; weak HEV, which supports the propulsion through an electric motor; strong HEV, which operates the electric motor when accelerating and maintaining vehicle speed; and plug-in HEVs, which are basically operated by an electric motor but also have an engine operated only when the vehicle needs power beyond the battery's capacity. The presence of the electric powertrain is intended to achieve either better fuel economy or better performance than a conventional vehicle.

Battery electric vehicle and fuel-cell electric vehicle are classified as electric vehicles using pure electrical power. A BEV is a type of EV that uses chemical energy stored in a rechargeable battery pack. BEVs use electric motors and motor controllers instead of ICEs for propulsion. BEVs derive all their power from their battery packs and, thus, have no internal combustion engine, fuel cell, or fuel tank. An FCEV is a type of hydrogen-powered vehicle that uses a fuel cell to produce electricity, powering its onboard electric motor.

### 2.1.5 Market Growth

According to the IEA 2010, the sales volume of EFVs is forecasted to gradually increase because of increasingly stringent pollution regulations, as shown in Figure 2.5. It is estimated that the United States will dominate about 62.9% of the EFV market, and Japan and the EU are forecasted to account for 21.9% and 14.5%, respectively. Hybrid vehicles dominate that initial share of EFVs and are led by Japanese suppliers with a 90% market share. Market forecasts predict that the EFV market share will continue to increase to fulfill $CO_2$ regulations. The global sales volumes of HEVs, PHEVs, and BEVs are expected to reach 4.9%, 1.9%, and 0.9%, respectively, and the global sales volume of PHEV and the total volume of EFV will reach about 19.6% in 2020 [2-1].

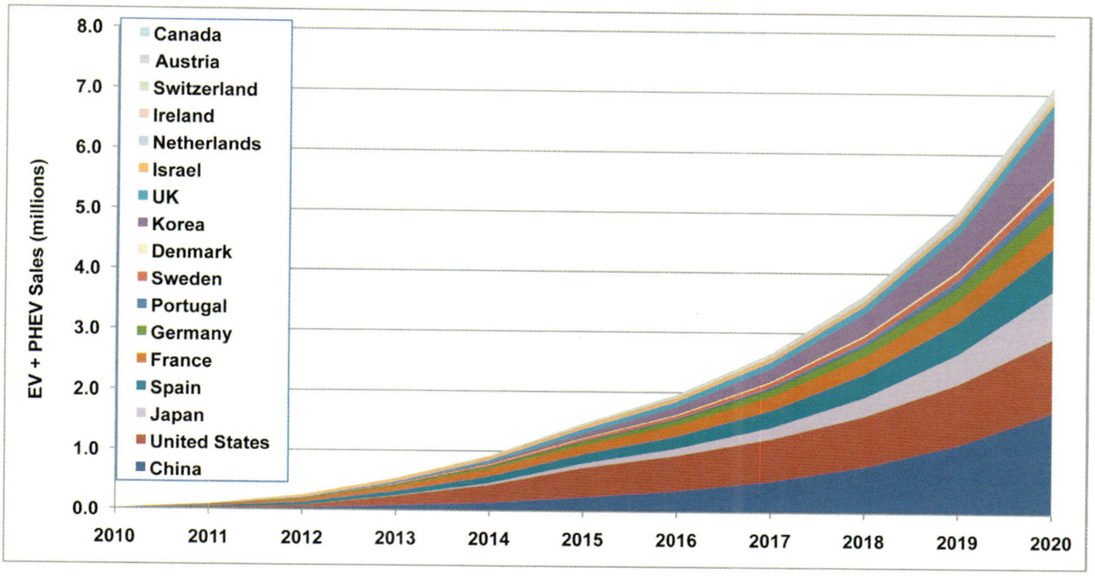


**Figure 2.5** National EV/PHEV sales target if national target year growth rates extend to 2020 (© 2011 OECD/IEA. Reprinted, with permission, from [2-1], IEA Publishing. License http://www.iea.org/t&c/).

### *2.1.5.1 U.S. Market Forecasting*

The forecast for EFVs based on the market demand in the U.S. for 2009 was 281 thousand vehicles, which is 2.7% of the total recorded vehicle demand, as shown in Table 2.3 [2-10].

**Table 2.3 Total U.S. EFV Market Demand Forecast for 2010–2020 (2009)**

| Type | 2009 | | 2010 | | 2015 | | 2020 | |
|---|---|---|---|---|---|---|---|---|
| | Unit | MS | Unit | MS | Unit | MS | Unit | MS |
| *HEV* | *281* | *2.7%* | *414* | *3.6%* | *1,300* | *8.0%* | *1,914* | *11.0%* |
| *PHEV* | *0* | *0.0%* | *23* | *0.2%* | *488* | *3.0%* | *1,218* | *7.0%* |
| *BEV* | *0* | *0.0%* | *12* | *0.1%* | *244* | *1.5%* | *696* | *4.0%* |
| *Total* | *281* | *2.7%* | *449* | *3.9%* | *2,032* | *12.5%* | *3,829* | *22.0%* |

Unit: thousands, MS: Market Share

To satisfy the $CO_2$ reduction target for 2020, the share of EFVs should be increased by 62%; the portion of pure electric vehicles such as PHEV and BEV is expected to be about 11% of this share [2-10]. In the U.S. market, the share of mid or large vehicles is greater than in the EU or Japanese market, so $CO_2$ emissions are higher than for these regions. To fulfill the 152 g/km $CO_2$ emission target by the year 2016, focus must be increased on existing mild and full HEVs. Then, PHEVs and BEVs can be gradually introduced to meet the 130 g/km of $CO_2$ emission target by the year 2020.

### 2.1.5.2 EU Market Forecasting

The market demand for EFVs in the EU in 2009 was forecasted as 48.4 million vehicles, which is 3.6% of the total demand, as shown in Table 2.4 [2-10].

**Table 2.4 Total EU Market Demand Forecast 2010–2020 (2009)**

| | 2009 | | 2010 | | 2015 | | 2020 | |
|---|---|---|---|---|---|---|---|---|
| Type | Unit | MS | Unit | MS | Unit | MS | Unit | MS |
| *HEV* | *75* | *0.6%* | *140* | *1.1%* | *1,203* | *7.0%* | *2,779* | *15.0%* |
| *PHEV* | *0* | *0.0%* | *0* | *0.0%* | *429* | *2.5%* | *2,223* | *12.0%* |
| *BEV* | *0* | *0.0%* | *0* | *0.0%* | *172* | *1.0%* | *1,482* | *8.0%* |
| *Total* | *75* | *3.6%* | *140* | *1.1%* | *1,804* | *10.5%* | *6,484* | *35.0%* |

Unit: thousands, MS: Market Share

However, diesel engines, especially in light vehicles, take up more than half of the total, so EFVs are not expected to have considerable impact in the EU market. To satisfy the targets of the $CO_2$ regulations for 2015 and 2020, volumes of EFVs would be increased to 10.5% and 35%, respectively. The portions of BEVs and PHEVs are expected to make up 8.0% and 12% of the market by 2020, respectively. The EU has the most stringent $CO_2$ regulations compared to the other region, so the diffusion of EFV is expected to be faster than in other regions.

### 2.1.5.3 Japanese Market Forecasting

The market demand of EFVs in Japan is recorded at about 39 million vehicles, which was 8.8% of the total demand in 2009, as shown in Table 2.5 [2-10]. Although the Japanese market has high demand for light vehicles, EFVs have increased based on taxes levied to favor hybrid vehicles. The quantity of EFVs necessary for meeting the $CO_2$ emission targets for 2015 and 2020 are estimated to be 25.4% and 69%, respectively. BEVs and PHEVs are estimated to occupy 1.9% and 12.0% of the market in 2015 and 2020, respectively. Japan has regulations similar to Europe, and major global OEMs such as Toyota and Honda have interest in EVs. Thus, Japan is currently leading HEV technologies. In addition, because Japan has legislated micro mobility for vulnerable groups in transportation, traffic congestion, and parking lot shortages, various EV technologies are expected to be diffused and developed.

| Table 2.5 Total Japan Market Forecast 2010–2015 (2009) | | | | | | | | |
|---|---|---|---|---|---|---|---|---|
| | 2009 | | 2010 | | 2015 | | 2020 | |
| Type | Unit | MS | Unit | MS | Unit | MS | Unit | MS |
| *HEV* | *387* | *8.8%* | *432* | *9.7%* | *706* | *14.0%* | *1,355* | *27.0%* |
| *PHEV* | *0* | *0.0%* | *0* | *0.0%* | *50* | *1.0%* | *351* | *7.0%* |
| *BEV* | *0* | *0.0%* | *9* | *0.2%* | *45* | *0.9%* | *251* | *5.0%* |
| *Total* | *387* | *8.8%* | *441* | *9.9%* | *801* | *15.9%* | *1,956* | *39.0%* |

Unit: thousands, MS: Market Share

### 2.1.5.4 Chinese Market Forecasting

The market demand for EFVs in China is estimated to be about 0.4 million vehicles, which is 0.4% of the total demand, as revealed in Table 2.6. Although EFV utilization is gradually increasing, it is not sufficient from the perspective of the total demand. The required market shares of EFVs to meet $CO_2$ emission targets for 2015 and 2020 are estimated to be 7.0% and 20%, respectively. The market share of BEVs and PHEVs needed to meet the emission targets by 2015 and 2020 are 4% and 15%, respectively [2-10]. Because China is an emerging market, less stringent environmental regulations are in place compared to most advanced countries. However, investment in EFVs initiated by the government is increasing, to lead the global automotive market. Technologically specialized OEMs such as BYD, an advanced battery technology company, are increasing and as a result, the Chinese market is predicted to become dominated by pure electric vehicles rather than HEVs.

| Table 2.6 Total Chinese Market Forecast 2010–2020 (2009) | | | | | | | | |
|---|---|---|---|---|---|---|---|---|
| | 2009 | | 2010 | | 2015 | | 2020 | |
| Type | Unit | MS | Unit | MS | Unit | MS | Unit | MS |
| *HEV* | *31* | *0.3%* | *124* | *1.1%* | *446* | *3.0%* | *941* | *5.0%* |
| *PHEV* | *0* | *0.0%* | *45* | *0.4%* | *297* | *2.0%* | *1,129* | *6.0%* |
| *BEV* | *10* | *0.1%* | *34* | *0.3%* | *297* | *2.0%* | *1,694* | *9.0%* |
| *Total* | *41* | *0.4%* | *202* | *1.8%* | *1,040* | *7.0%* | *3,765* | *20.0%* |

Unit: thousands, MS: Market Share

## 2.2 Intelligent Transportation System

Intelligent transportation systems (ITS) are advanced applications that, without embodying intelligence as such, aim to provide innovative services relating to different modes of transport and traffic management. They also enable users to be better informed and make safer, more coordinated, and "smarter" use of transport networks [2-10]. Although ITS may refer to all modes of transport, EU Directive 2010/40/EU [2-11], which details the framework for the deployment of intelligent transport systems in the field of road transport and for interfaces with other modes of transport, defines ITS as systems in which information and communication technologies are applied in the field of road transport, including infrastructure, vehicles, and users, and in traffic management and mobility management, as well as for interfaces with other modes of transport. The global ITS market is expected to gradually increase and the U.S., EU, and Japan are predicted to lead this market with tremendous investment. The underlying concept of ITS development can be described as in Figure 2.6.

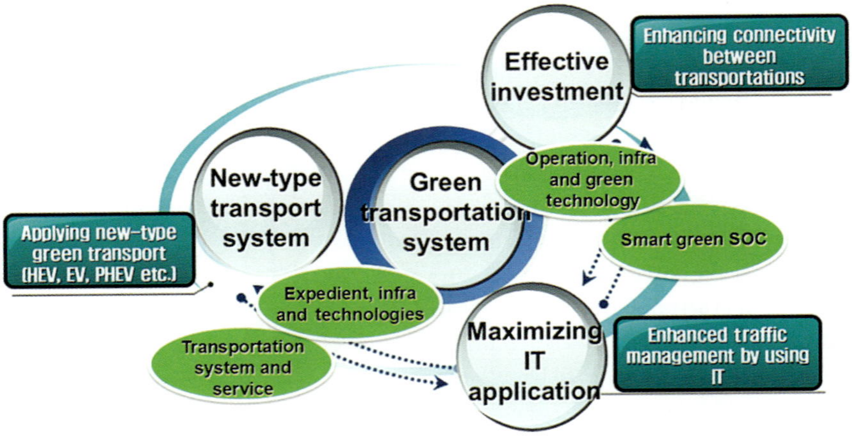


**Figure 2.6** Concept of ITS development.

The general purpose of ITS can be described in six sections: traffic congestion reduction through improving transportation efficiency and road capacity; driver's mobility, convenience, and safety enhancement through improving traveler services; transportation system stability improvement; advanced technological approaches for future transportation systems; environmental impact and costs reduction through efficient energy consumption and air pollution reduction; and vehicle and related industry development.

ITS technology is being researched and developed throughout the world to achieve those goals, and it is converging with transportation and communication technologies. ITS incorporates future mobility technologies, such as vehicle platooning and autonomous driving, and it requires a huge amount of data for the driving environment. Thus, the technologies related to ITS have to be developed simultaneously with vehicle technology and communication technology.

## 2.2.1 Intelligent Vehicle Technology

The purpose of intelligent vehicle technology is to minimize traffic accidents and congestion by applying electronic equipment in the vehicles. Intelligent vehicle technology level can be described by degree of automated function. As shown in Figure

2.7, intelligent vehicle technology can be classified into four categories: driver warning systems, active safety controls, V2X communication, and autonomous driving.

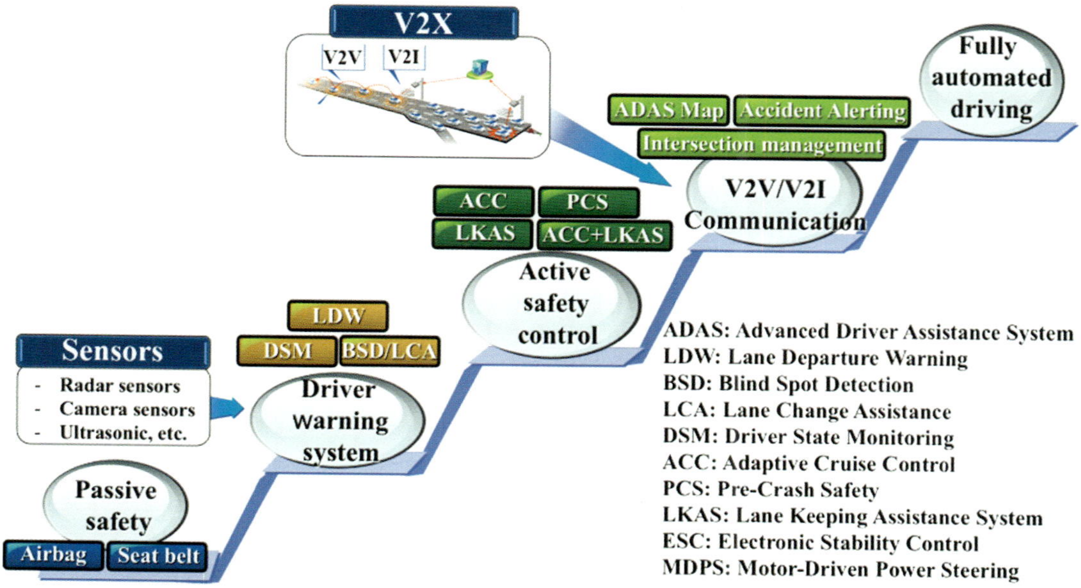


**Figure 2.7** Development process of intelligent vehicle technology.

Driver warning systems are technology that provides alerts of hazardous circumstances that might occur while driving. Such systems only provide alerts to the driver, so they cannot prevent accidents completely. To reduce accidents more thoroughly, active safety control technologies execute safety functions without the driver's control, so they can prevent accidents via electronic equipment. Active safety technologies depend on the electronic systems equipped in an individual vehicle, so they can prevent accidents before they occur. However, active safety technologies cannot contribute to road capacity enhancement because active safety functions operate only for safety by electronic or mechanical units that are equipped in the vehicle. Therefore, to manage traffic congestion safely and efficiently, it is necessary to converge other technologies such as information technology (IT), and the vehicle should communicate with the other vehicles or infrastructures.

To achieve safe and efficient traffic operation, V2X technology is being developed. V2X is a technology that connects a vehicle with other objects, such as infrastructure, nomadic devices, or other vehicles, through communication technologies. V2X technology can use external data related to vehicle driving, so it can provide not only convenient driving but also efficient traffic operation. The highest level of vehicle technology is autonomous driving, which can drive a vehicle without human control so that the vehicle moves to its destination automatically. Consequently, drivers can spend their time on entertainment, work, or rest, arriving at their destinations without incident. Autonomous driving includes vehicle platooning, autonomous parking, etc.

## 2.2.2 Cooperative ITS

Cooperative ITS (C-ITS) is a V2X technology that communicates with other objects such as infrastructure and vehicles. The difference between ITS and C-ITS can be described by inter-operability: the way in which objects related to transportation are communicated and controlled. Conventional ITS can be operated for collecting information about transportation without delivering information to other vehicles. The information collected by infrastructure is delivered to a traffic management center so that the traffic information can be used for measuring traffic conditions. In addition, because ITS does not include communication between vehicles, the only vehicle technologies that can be included are those active safety technologies equipped into the vehicle. The fundamental concept of C-ITS is described in Figure 2.8.

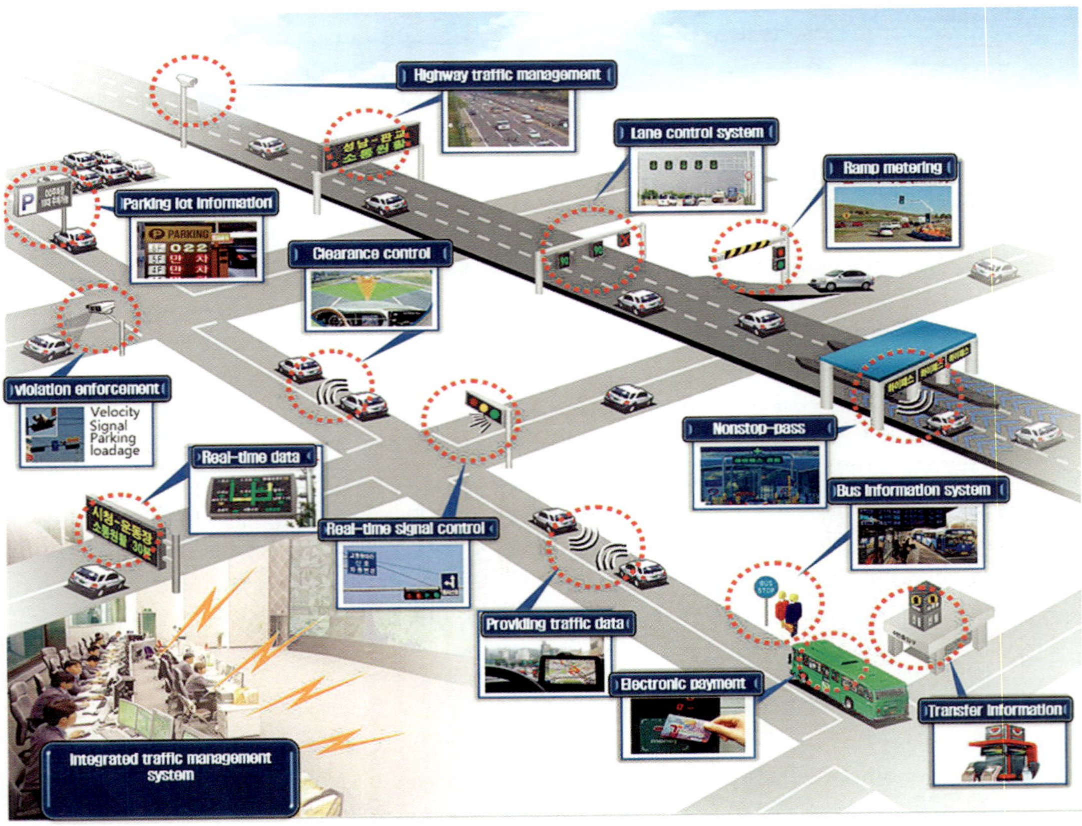


**Figure 2.8** Concept of the C-ITS environment.

The C-ITS is operated based on the communication between all transportation system elements such as the vehicle and infrastructure. This technology is called V2X technology and can be classified according to which object is communicating with the vehicle. If the vehicle is communicating with the infrastructure, it classified as V2I. The purpose of V2I communication is to use external information about current traffic conditions such as congestion and weather so that the vehicle can drive on the road efficiently using these data. If a vehicle communicates with other vehicles, it is classified as V2V. The purpose of V2V is to maintain safe and efficient driving. This means that vehicle information related to movement, such as lane changes, acceleration, and deceleration, is shared with other vehicles so they can derive appropriate movement patterns. For example, if one vehicle will change lanes, a signal will be delivered to adjacent vehicles so the vehicles are operated in such a way as to allow the vehicle to perform the maneuver safely.

## 2.3 Electric Vehicle Charging Systems

EV charging system technologies are key elements for diffusing EV. EV charging systems are influenced by several factors, including charging technology, policy, and applicability of existing technology. EV charging systems are classified roughly into three categories: conductive charging, battery exchange, and wireless charging. These charging systems have advantages and drawbacks that must be considered when evaluating their potential merits for application in future transportation. Wireless charging systems were described in chapter 1, so they will not be explained again in this chapter.

A conductive charging system is a technology that can charge a BEV or PHEV through direct connection by electrical conduction, as shown in Figure 2.9. Conductive charging systems receive electrical power from the grid so that they can supply electrical power to vehicles through physical connections with related equipment such as connectors, power systems, and grounding.

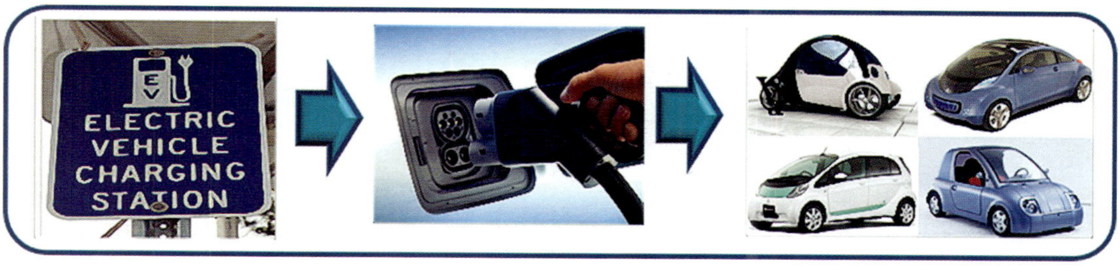


**Figure 2.9** Plug-in type conductive charging.

Charging time differs depending on the charging type used. As an example, Table 2.7 shows three charging levels of the SAE J1772 charger [2-12]. Ac level 1, which is low-speed charging, can use a physical connection installed in the home. Level 1 charging time is about five hours. Ac level 2, which is medium-speed charging, requires an additional power system when compared to level 1. Although the ac level 2 charging needs an additional system, it can provide faster charging than ac level 1. The ac level 2 charging time is about one or two hours. Dc fast charge is the charging system that can provide the fastest charging among of the available systems. Although the dc fast charge is the fastest charging system, it requires the installation of additional separate power system components above a general home power system. This means that the application of dc fast charge is complicated because of the existing power system infrastructure, which is why dc charge systems are not the universal charging type yet.

**Table 2.7 Types of SAE J1772 chargers (reprinted from SAEJ1772, February 2012)**

| Charging Method SAE J1772 | EVSE Voltage (Vrms) | EVSE Current (Arms) | Branch Circuit Rating (Circuit Breaker, Amps) | Charging Rate (miles range/h of charging) |
|---|---|---|---|---|
| *ac level 1* | *120, 1-phase* | *12* | *15* | *2 to 5* |
| | *120, 1-phase* | *16* | *20* | |
| *ac level 2* | *208 to 240, 1-phase* | *<80* | *NEC article 625* | *10 to 20* |
| *dc fast charge* | | | | *60 to 80* |

Battery exchange systems exchange an almost discharged battery equipped in a vehicle with a fully charged one through infrastructure called a quick change machine (QCM) in about 5 minutes, as shown in Figure 2.10. The main advantage of the battery exchange system is that it is time-efficient. In addition, it is not necessary to buy a battery because the batteries would be circulated among the battery exchange system customers.

**Figure 2.10** Various battery exchange stations.

However, different types of battery change infrastructures are required because of various vehicle types, such as buses, trucks, and sedans, but this has some drawbacks. The charge system can be used inefficiently with respect to batteries and vehicles. In addition, because the cost for installing a charging system is greater than that for a plug-in conductive charge, the number of battery exchange stations would be less than plug-in charging stations with the same budget.

### 2.3.1 Market Trend

Developments of charging systems are gradually increasing throughout the world based on governmental support, but technological advancement is not yet sufficient, especially as related to charging time. To resolve this technological limitation, research and development activities are underway worldwide.

The U.S. government is currently developing technology that combines a smart grid with EFV-charging infrastructure. Vehicle designs for efficient energy use are constantly being proposed. U.S. EFV promotion is initiated through "The American Recovery and Reinvestment Act of 2009," and $2.4 billion have been allocated to the development and promotion to EFVs and their related infrastructures. Relying on governmental support, various EV-charging systems developed by global companies have been commercialized in the United States, as shown in Figure 2.11.

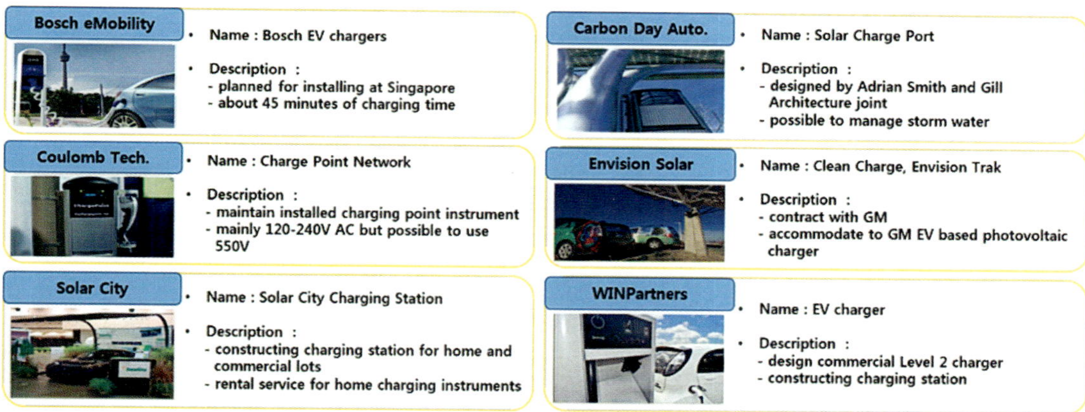


**Figure 2.11** EV charging business in the United States.

EU countries are also focusing on the EV charging infrastructure with governmental support. The UK government created the Office for Low Emission Vehicles (OLEV) to promote the adoption of Electric Vehicles in the UK in 2009. Based on this promotion, the Department for Regional Development (DRD) and Department of the Environment (DOE) of the UK have supported a project called "Plugged in Places (PIP)" that installs charging stations. This project mainly offers to install EV recharging points with match-funding to consortia of businesses and public sector partners. The UK government is supporting PIP for eight areas: east England, Greater Manchester, London, Midlands,

Milton Keynes, northeast England, Northern Ireland, and Scotland. Currently, over 400 charge points have been established, and about 65% of them are publicly accessible [2-13]. In addition, the Transportation Research Laboratory (TRL) of the UK has announced that other EU countries are also planning for the implementation of a charging system. Denmark is implementing a demonstration project supported by the government for charging infrastructure with a consortium called "Edison" involving various companies such as IBM and Siemens.

The Japanese government is not only implementing charging infrastructure but also trying to create a global standard charging technology called "CHAdeMO." Global EV-charging technology is being discussed among advanced countries. For this reason, the Japanese government is focusing on the development and diffusion of its charging technologies. The budget planning for EFV and charging technologies introduced by the Japanese government is about $1.4 million for 2011, and the government is planning for a gradual increase of technology diffusion so that the budget would be gradually increased.

Most advanced countries are trying to lead the EV-charging market by implementing a diffusion strategy with standardization with governmental support. The charging system market is being gradually increased along with the EFV market. Technological competitiveness will be the most important factor for the standardization and leading of the global market.

### 2.3.2 Challenges

The challenges of EFV-charging systems are classified three categories: technological limitations, social acceptance, and political limitations. These categories are interrelated.

The technological challenge is to provide high-quality technology at low cost. The current EFV-charging infrastructures have tradeoffs between charging time and cost. Slow-charging systems can be used inexpensively but are time-consuming. On the contrary, fast-charging systems take little time but require high installation cost. In addition, technological standards are not yet established. The technological standards incorporate specifications and safety tests. Absence of charging system standards brings risks for manufacturers such as redesigning and remanufacturing of their own products. These risks will cause delays in the market penetration of EFVs because of insufficient infrastructure. Technological limitations of EV charging brings negative effects for social acceptance and EV diffusion because customers cannot ensure safety of EV charging. In addition, insufficient infrastructures of EV charging bring inconvenience of EV utilization.

Even if the technological limitations can be resolved, commercialization has other limitations. Current charging infrastructure installations are expensive because the implementation of charging systems is currently in its initial stages, and EV-charging systems are not yet universal technologies, so they have insufficient supporting bases such as labor and supply chains. In addition, EV charging infrastructures are required to cooperate with existing electric power grids that supply electrical power to the consumer. Insufficient charging infrastructures will result in EV implementation delays, making their earning potential through charging system businesses uncertain. To achieve successful charging system diffusion, governmental supports and interoperability with existing infrastructures should be considered.

Most advanced countries try to overcome the aforementioned technological and social limitations through policy instruments such as subsidies for EFV purchase, tax benefits, and governmental support for technological development. However, promotion policies for EFV and their charging technologies have limitations because current policies are insufficient to implement EV-charging businesses. This means that the EV-charging business is still quite risky due to insufficient EV users. Installation of charging infrastructure has an intimate relationship with existing industries such as parking lots, power systems, and electrical generation. However, the policy for cooperative development is insufficient. Consequently, installing charging infrastructure might also be difficult due to complicated legal processes.

For the deployment of charging infrastructure to advance, appropriate policy strategies to manage these factors must be put in place. Structurally appropriate charging type designs according to the place have to be evaluated, as well as do the advantages and drawbacks of each type of charging system. The direction of charging system diffusion also has to be determined according to the future transportation environment, which is incorporated with lifestyle, future transportation systems, and technologies.

## 2.4 WPT Application to EV

While most car trips are under 50 miles in global industrialized countries, consumers can hardly be satisfied with the current 100-mile range of typical EVs. Without any technological breakthrough in onboard energy storage systems, WPT can provide a significant opportunity in charging EVs. Wireless charging test beds are already in place and even commercialized in various regions, including Korea, the United States, and EU countries. Static wireless EV charging will become popularized when international standards SAE J2954, IEC 61980, and ISO 19363 are completed within approximately a two-year time frame. The current working modules of technical standards are focusing on Level I or Level II slow charging, up to a range of 7.7 kW.

It should be noted that several wireless charging projects are funded by EU and its member countries; they include UNPLUGGED, ZeEUS, and FABRIC. The "UNPLUGGED: Wireless Charging of Electric Vehicles—European Commission co-funded Project," led by FKA in Germany and 17 partners from 7 countries, aims to investigate how the application of wireless charging of EVs in urban environments improves the convenience and sustainability of car-based mobility. In particular, it will examine the feasibility of how smart wireless en-route charging infrastructure can facilitate full EV integration in the urban transport system while improving customer acceptance and perceived practicality. The "ZeEUS: Zero Emission Urban bus system," coordinated by UITP with a consortium of 40 partners from 10 countries, is working on a 42-month demonstration project aimed at extending the fully electric solution to a wider part of the urban bus network. The electric buses being deployed in the UK will use static wireless charging at bus terminals during the day. The "FABRIC: Feasibility Analysis and Development of On-Road Charging Solutions for Future Electric Vehicles" is also a European Commission co-funded project, of which the objective is to assess the potential and feasibility of a more extensive integration of electric vehicles in the mobility and transportation system, with a primary emphasis on dynamic wireless charging or charging while driving.

To truly demonstrate WPT technology, it is necessary to build an electric vehicle with a power receiving mechanism installed. The power electric system architecture of such a vehicle as well as the major function controllers within an electric vehicle are shown in Figure 2.12. The regulator in Figure 2.12 is for power flow management from the road surface to the vehicle. The motor control unit (MCU) and battery management system (BMS) also communicate with each other and with the main vehicle controller unit (VCU), including the main circuit breaker assembly of the power distribution unit (PDU). In addition to the battery, a control mechanism for a super capacitor is also included. The super capacitor management system (CMS) manages bi-directional dc-dc converter (BDDC) operation, and BDDC transfer electrical power to the PDU.

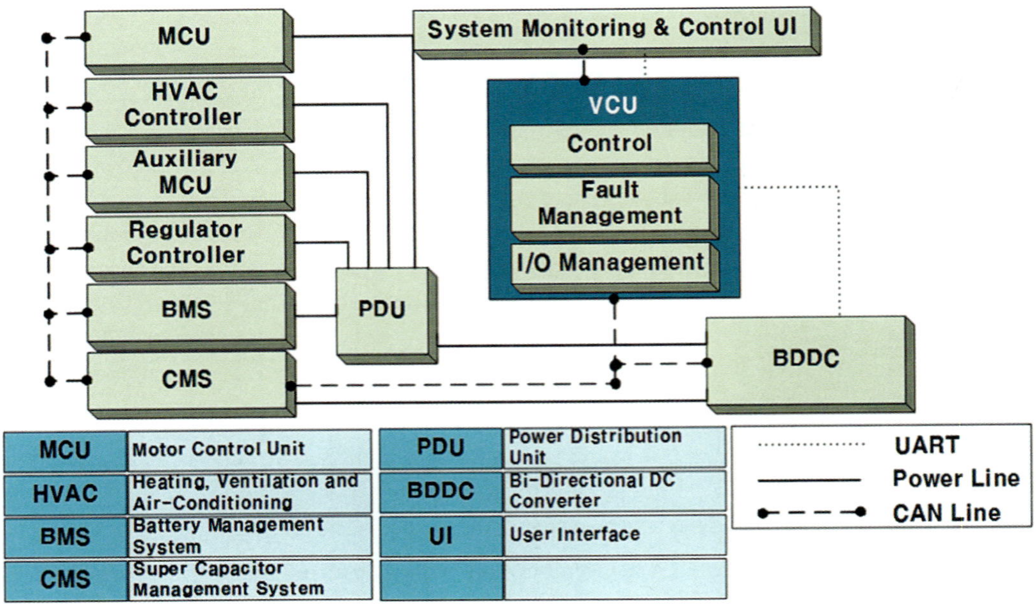


**Figure 2.12** OLEV vehicle system architecture in a control network.

Power cables inserted in fiber reinforced plastic (FRP) tubes are buried in concrete laid on the rib structure of assembled ferrite-core frames. The FRP tubes are used to protect the cable while providing extra strength to the concrete. The structural strength has been verified under a series of repeated loading conditions following standard road-construction test procedures and considering environmental conditions such as seasonal temperature changes and variations in humidity. The targeted life expectancy of the road construction was set at 20 years of continuous use with proper procedures for maintenance and a typical maintenance cost of about 8% of the initial investment.

With wireless charging with WPT during vehicle operation, the battery energy storage capacity can be smaller [depending on the vehicle travel pattern and distance, as well as the power supply infrastructure-installed rate (PIR) along the traveled route]. The process for optimizing PIR for closed-circuit-vehicle travel routes has been described well [2-14]. In general, about a quarter of the size of a typical EV battery is required with a PIR of about 20%. The PIR can be considered zero for a pure battery-operated EV (a plug-in electric vehicle, PEV). Thus, when designing OLEV, the battery capacity and PIR will be design variables for optimization in relation to installation cost and required vehicle performance, while they are fixed design variables for PEVs and particularly for closed-circuit-vehicle operation.

Additional components of an EV with OLEV technology are a set of pick-up modules incorporated with capacitor banks, a regulator, and necessary communication and control integration within a vehicle. While converting from a typical diesel-engine operated bus (12,000 kg curb weight), the weight was reduced by about 1000 kg. Compared with a pure electrical bus, the reduced weight can be estimated to be about 25% of the typically required battery capacity.

### 2.4.1 Efficiency of WPT

The efficiency for an electrical or mechanical system can be generally defined as the ratio of the output and input power or energy ($\eta_0 = P_{0,out}/P_{0,in}$) of the system, as shown in Figure 2.13. For wireless power transmission, the efficiency when the power is transmitted through the air is used. While this provides a direct measurement of the efficiency of the wireless power transmission through the air between the power supply and receiver, two issues may arise.

1. It is not easy or straightforward to collectively measure the surface integral of the magnetic flux of both surfaces.
2. Measurements made through the air do not provide an effective real figure for wireless power transfer from the system point of view.

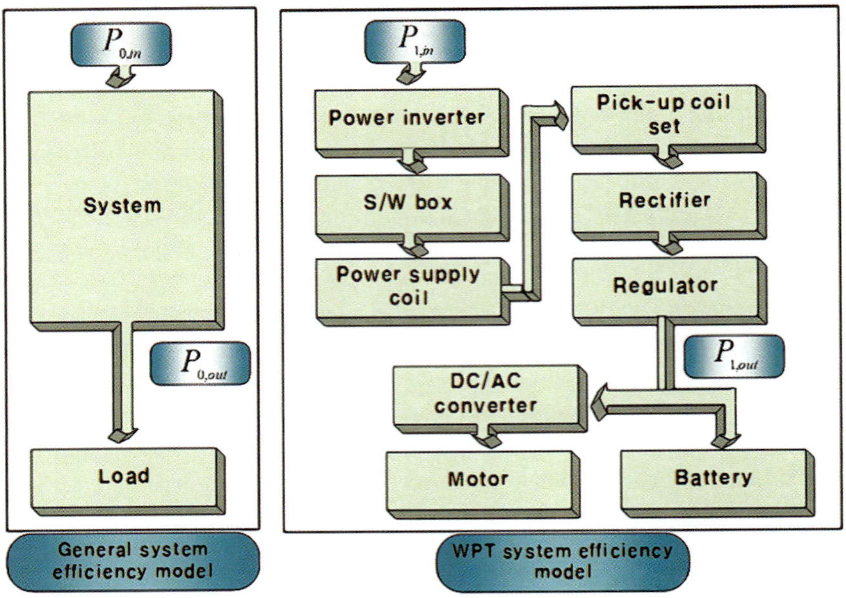


**Figure 2.13** Power transfer efficiency definition for a vehicle with wireless power transfer.

Thus, an effective measure of the WPT system efficiency can be defined as for the system shown in Figure 2.13. The electrical power is supplied from the electric power grids, and it is transferred to the power inverter to operate the wireless power transfer system. Received power from the electric power grids is transferred to the power supply coil and is converted to a magnetic field. The pick-up coil set receives electrical power from the power supply coil through the magnetic field, and the received electrical power is regulated and rectified at the rectifier and regulator. Regulated power is transferred to the dc-ac converter or battery.

Whether power is transferred to the dc-ac converter or transferred to the battery is determined by the instantaneous power required by the drive motor during acceleration or deceleration and by the charging rates of different types of batteries and SOC status. The efficiency referred to in this section follows this definition of efficiency: the maximum efficiency is achieved when the maximum power is transferred. This is because power losses through the power electronic components are relatively constant during operation.

Considering the magnetic coupling between the two resonant coils, the parameters affecting the transfer-effectiveness are the resonance response at the resonant operating frequency, the mutual inductance and coupling factor of the two circuits, the power losses involved within the power electronics (i.e., inverter, pick-up coils, capacitor banks, and regulator), and the skin effect of the power cables and the leakage field of the magnetic field [2-15], [2-16], [2-17]. The efficiency is also strongly dependent on the power capacity because the power losses remain relatively constant regardless of the maximum power transfer rate. In one example, when 70 kW of power is supplied, the measured power losses are 12% in the inverter, 12.5% in the power cable and pick-up coils, and about 5% in the regulator. Since 50 kW was transferred in the earlier design, the measured efficiency was about 71% in that case.

If the vehicle operating mode is considered from the perspective of the transmission effectiveness measure, the parameters involve many other factors (e.g., misaligned vehicle position with the center of the road; electric load depending on the instantaneous power required by the vehicle; operating temperatures of the battery cells; battery charge and discharge characteristics; and vehicle operating conditions such as acceleration, deceleration, and speed). At ±20 cm misalignment locations, the transmitted power was about half that of the perfectly aligned condition; however, the efficiency was the same as for the case of perfect alignment. With ±30 cm misalignment, the magnetic coupling between the two resonant circuits was lost. This means that the power delivery environment from the power supply module to the pick-up module must be considered in the OLEV system.

## 2.5 Implications of Wireless Charging in Future Transportation

### 2.5.1 Efficient Power Management

Currently, most electrical energy is supplied from thermal or nuclear power generation. These power generation methods have issues such as limited resource availability, nuclear waste, and $CO_2$ emissions. For this reason, currently operated EFVs are not completely eco-friendly because of the power generation resources. The electrical energy can be generated by various types of power generation including renewable energies such as photovoltaic, biomass, and wind, which are currently being advanced for preserving the environment. In addition, most transportation takes place during the daytime, so the electrical energy might be deficient because numerous machines use electrical energy during the daytime.

The vehicle-to-grid (V2G) concept, which uses hybrid vehicles or pure EVs to store and supply energy back to the grid, is gaining more and more popularity as hybrids and EVs are considered to be an indispensable component in both "living and mobility" and sustainable living in the near future [2-18], [2-19]. To resolve the power shortage, intelligent power systems have to be used for controlling electrical power balances. Next-generation power systems such as smart grids can be applied not only for efficient power distribution but also for converging with renewable energy systems.

Irrespective of whether an EV or a fleet of EVs is used solely for medium-scale energy storage or microscale residential use, as in the case of "living and mobility" [2-19], there lies a challenge of the charging and retrieving (discharging) energy. Consequently, techniques for charging and discharging EVs, with emphasis on simplicity, low cost, convenience, high efficiency, and flexibility, have become one of the main focuses of current research in both industrial and academic communities whose fields of interests are in V2G and sustainable living [2-20], [2-21]. Contactless or wireless charging techniques are emerging as a viable choice as they meet most of the aforementioned attributes [2-15], [2-16], [2-17]. If a WPT system were converged with IT, environmentally friendly transportation systems would be efficient in averting power shortages.

### 2.5.2 Safety

Charging systems have two major areas of safety: electrical safety and electromagnetic field (EMF) issues for humans. Electrical safety indicates insulation of electrical systems and protection for people against electric shock. To assure electrical safety, electrical safety regulations are necessary. In the case of a conductive charging system, the charging connectors have to be connected by a human when the charging system supplies electrical power to the vehicle. A danger is that the user could be electrically shocked. The WPT can be used against these hazards because WPT charging is operated without human interaction. This means that humans do not need to be around the charger because the WPT is operated not only when the vehicle operation is stopped, such as parking, but also while driving.

EMF issues indicate the need for EMF shielding against human risk. Electrical power flow generates an EMF that might be harmful for human health. The reference values for levels of occupational and general public exposure of humans to time-varying electric and magnetic fields in unperturbed values are defined according to the International Commission for Non-Ionizing Radiation Protection (ICNIRP) [2-22]. The suggested reference value at 20 kHz for the general public is 27.0 μT (e.g., 270 mG in the region of 3 kHz to 10 MHz), while the reference value for occupational exposure is 100 μT (1000 mG in the same frequency range) [2-15], [2-16], [2-17]. For the OLEV application, which uses a dynamic WPT, it is suggested to measure the EMF level at 20 kHz, 30 cm from the outer body line of the vehicle, taking the average of three measures (at 0.5, 1.0, and 1.5 m off the ground surface), while the wireless power transfer is in action on a stationary vehicle. The measured value met the reference values suggested by ICNIRP for any condition of operating power transmission. Consequently, WPT can provide safer charging than conductive charging.

### 2.5.3 Convenience

With the wireless charging technology, the charging infrastructure can be more intelligently applied by developing the IT convergence concept linked together with the smart grid development. Figure 2.14 demonstrates one possible scenario of the smart charging infrastructure in the near future. It shows the accessibility for vehicle charging wherever vehicles are, and the vehicle can be charged so that it can maintain sufficient power for driving without human control [2-17]. It will provide more efficient power consumption and convenient transport operation.

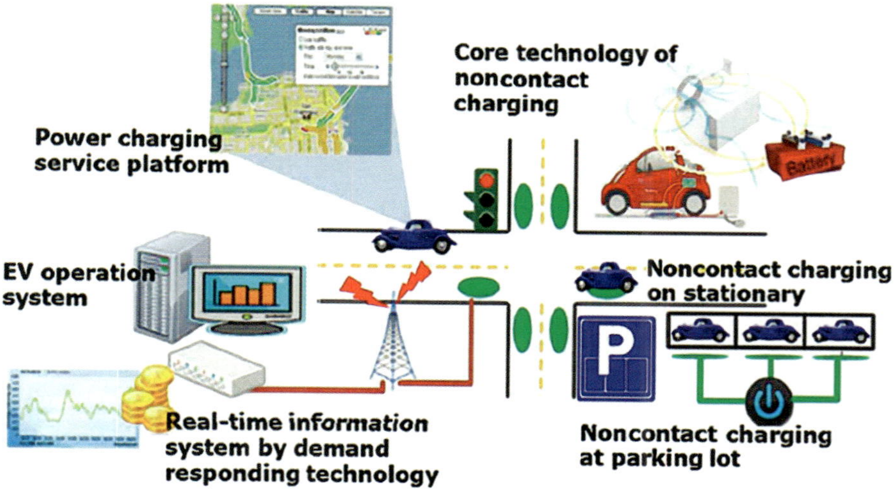


**Figure 2.14** An example of a smart charging infrastructure applying static and dynamic wireless charging technology (Reprinted ©2013 from [2-16] with permission from IOS Press).

### 2.5.4 $CO_2$ Reduction

The estimates of EV and PHEV sales used in this roadmap are based on achieving the Blue Map scenario's 2050 $CO_2$ reduction targets, which can only be met with the enactment of aggressive policies. $CO_2$ reductions also depend heavily on changes in electricity generation. Blue Map targets require nearly full decarburization of electricity generation around the world by 2050. As shown in Figure 2.15, the $CO_2$ intensity of electricity generation in the Blue Map scenario drops steadily over time until, by 2050, all regions have nearly decarbonized their electricity. This steady decrease is an important assumption; if the achievement of low $CO_2$ electricity generation around the world does not occur in the 2030 to 2050 timeframe, the $CO_2$ benefits of EVs and PHEVs will be much lower. The IEA is also developing roadmaps on achieving Blue Map electricity $CO_2$ intensity targets.

For PHEVs, $CO_2$ reduction levels will depend on the proportion of miles driven using battery electricity from grid recharging in lieu of petroleum consumption from an ICE. While it will take time to understand the relationship between the PHEV driving ranges as a function of the battery capacity, it is likely that even a modest battery power range (e.g., 40 km) will enable many drivers to cut petroleum fuel use by 50% or more, as the battery will cover their first 40 km of driving per day. In countries where the average driving distances per day are relatively short (e.g., Japan), a higher percentage of driving distance is expected to be covered by battery power than in countries with longer average driving distances (e.g., the United States).

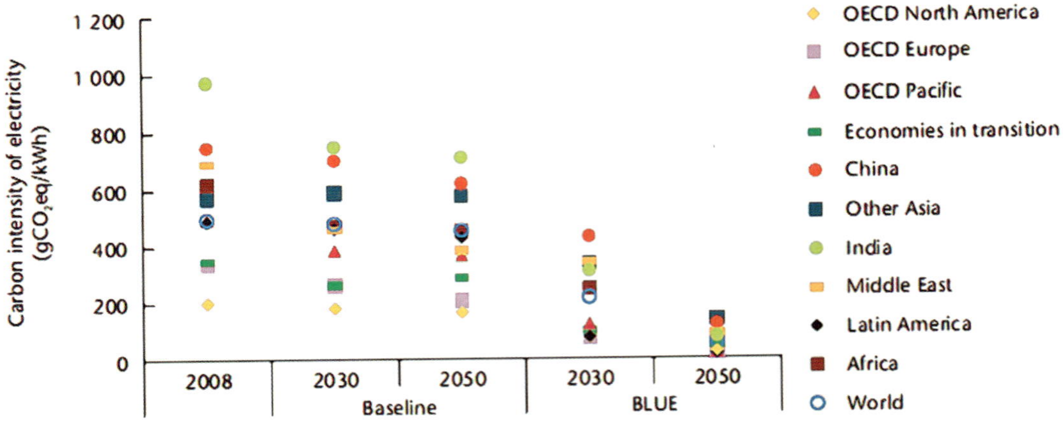


**Figure 2.15** $CO_2$ intensity of electricity generation by region, year, and scenario (© 2011 OECD/IEA. Reprinted, with permission, from [2-1], IEA Publishing. License http://www.iea.org/t&c/).

Overall, given the Blue Map scenario projections for the numbers of EVs and PHEVs deployed in the locations specified and assuming that these vehicles replace conventional gasoline vehicles (which themselves improve over time in the baseline), about 0.5 billion tons of $CO_2$ are projected to be saved per year worldwide in 2030, and about 2.5 billion tons are projected to be saved worldwide in 2050. With a Blue Map target of close to 500 million EVs on the road in 2050 and a $CO_2$ reduction of 2 tons (on a well-to-wheels basis) per vehicle per year compared to displaced gasoline ICE vehicles, EVs would provide about 1 billion tons of $CO_2$ reduction in that year. Approximately 800 million PHEVs would provide an additional 1.5 billion ton reduction.

## References


2-1. "Technology Road map: Electric and plug-in hybrid electric vehicles," IEA, June 2011.

2-2. J. Dargay, D. Gately, and M. Sommer, "Vehicle Ownership and Income Growth, Worldwide: 1960–2030," *The Energy Journal*, Vol. 28, No. 4, 2007, pp. 168–169.

2-3. A. Amekudzi, L. Thomas-Mobley, and C. Ross., "Transportation Planning and Infrastructure Delivery in 7 Major Cities and Megacities," *Transportation Research Record*, No. 8, pp. 17–23, Washington D.C., 2007.

2-4. C. Wessner, "FYI: What Is the Best Guess as to When the World's Oil Reserves Will Run Out?" *Popular Science*, February 1999, p. 76.

2-5. R. Kremlicka, "Mega trends in the automotive industry and its consequences for global players," A.T. Kearney, Nov. 2012.

2-6. "A global study on the passenger car powertrain market towards 2025," A.T. Kearney, May 2012.

2-7. R. Phaal, "Foresight Vehicle Technology Roadmap," University of Cambridge, Aug. 2002.

2-8. OECD, Factbook 2011–2012: Economic, Environmental and Social Statistics, 2012, website: http://www.oecd-ilibrary.org/economics/oecd-factbook-2011-2012_factbook-2011-en.

2-9. C. C. Chan, A. Bouscayrol, and K. Chen, "Electric, Hybrid, and Fuel-Cell Vehicles: Architectures and Modeling," *IEEE Transactions on Vehicular Technology*, Vol. 59, No. 2, pp. 589–598, Feb. 2010.

2-10. "Forecast on Hybrid and Electric Cars and Competitiveness Analysis of Korean Domestic Automotive OEMs," Hana Institute of Finance, 2009 (in Korean).

2-11. "EU directives: on the framework for the deployment of intelligent transport systems in the field of road transport and for interfaces with other modes of transport," July 2010.

2-12. "Electric Vehicle and Plug in Hybrid Electric Vehicle Conductive Charge Coupler," SAE J1772, October 2012.

2-13. Mayor of London, "An Electric Vehicle Delivery Plan for London," May 2009.

2-14. G. Erdmann, J. Kluge, P. Radtke, and H. Wallentowitz, "DRIVE-The Future of Automotive Power," McKinsey & Company, Inc., 2006.

2-15. I. Suh, "Application of SMFIR Technology to Urban Transportation," CIRP Design Conference, 2011, Daejeon, Korea.

2-16. J. Kim, J. Kim, S. Kong, H. Kim, I. Suh, N. P. Suh, D. H. Cho, and J. Kim, "Coil Design and Shielding Methods for a Magnetic Resonant Wireless Power Transfer System," Proceedings of the IEEE, 2013.

2-17. I. S. Suh, "Intelligent Wireless EV Fast Charging with SMFIR Technology," *Journal of Integrated Design and Process Science*, Vol. 15, No. 3, pp. 3–12, Apr. 2012.

2-18. B. Kramer, S. Chakraborty, and B. Kroposki, "A review of plug in vehicles and vehicle to grid capability," in Proc. IEEE Ind. Electron. Conf., 2008, pp. 2278–2283.

2-19. U. K. Madawala and H. Vincenz, "Living and mobility—A novel multipurpose in-house grid interface with plug-in hybrid blueangle," in Proc. IEEE, 2008, pp. 531–536.

2-20. U. K. Madawala and D. J. Thrimawithana, "A bidirectional inductive power interface for electric vehicles in V2G systems," IEEE Transactions on Industrial Electronics, Vol. 58, No. 10, pp. 4789–4796, Oct. 2011.

2-21. U. K. Madawala, D. J. Thrimawithana et al., "A two-way inductive power interface for single loads," in Proc. IEEE Ind. Technol. Conf., 2010, pp. 673–678.

2-22. "Guideline for Limiting Exposure to Time-Varying Electric and Magnetic Fields (1 Hz to 100 kHz)," International Commission on Non-Ionization Protection (ICNIRP), Health Physics Society, 2010.

# Chapter 3
## EV Charging Technology: Conductive and Wireless

This chapter introduces the distinction between conductive charging (cord set and plug) and wireless, or "magnetic resonance" charging, also known as inductive charging. The terms WPT and IPT (inductive power transfer), both of which are noncontact power transfers, are used synonymously.

### 3.1 Conductive Charging

All EVs receive their energy from the electric grid by plugging in and storing the electricity in an onboard battery pack. The charging rate is determined to a large extent by the capacity of the onboard battery, called an RESS (regenerative energy storage system), because it can be charged and discharged many times over its useful service life. Furthermore, because of manufacturer design and market requirements, the rating of the onboard charger (OBC) must be sufficient to recharge the EV battery in a suitable amount of time, for example 5 h or perhaps with dc fast charging in 30 min or less. With conductive charging, the access and delivery of utility-supplied electricity is performed via electric vehicle supply equipment (EVSE). The EVSE unit connects to the local utility grid and connects to the EV using a standard connector. Communications between EVSE and the EV are necessary to ensure that the charge rate matches the battery pack need and that it can be done safely. To the public, these EVSE units are referred to as "charging stations."

Standardization is essential in EV charging stations because EVSE may be produced by a variety of manufacturers, not to mention the lower tier suppliers who build the dedicated OBCs. As shown in Table 3.1, according to SAE International (SAE), the EVSE ratings fall into three categories: Level 1, Level 2, and Level 3. The associated charge connector pin is defined in Table 3.2.

**Table 3.1 EVSE Charger Ratings**

| Charging Method SAE J1772 | EVSE Voltage (Vrms) | EVSE Current (Arms) | Branch Circuit Rating (Circuit Breaker, Amps) | Charging Rate (miles range/h of charging) |
|---|---|---|---|---|
| *ac level 1* | *120 V ac, 1-phase* | *12* | *15* | *2 to 5* |
| | *120 V ac, 1-phase* | *16* | *20* | |
| *ac level 2* | *208 to 240, 1-phase* | *<80* | *NEC article 625* | *10 to 20* |
| *dc fast charge* | | | | *60 to 80* |

**Table 3.2 Charge connector pin**

| Pin # | Connector Function | Vehicle Receptacle Function | Description |
|---|---|---|---|
| *1* | *Grid power, L1* | *OBC L1* | *Grid connection level 1 and 2* |
| *2* | *Grid power, L2* | *OBC L2* | *Grid connection level 1 and 2* |
| *3* | *Equipment ground* | *Chassis ground* | *Connects EVSE ground to EV chassis* |
| *4* | *Control pilot* | *Control pilot* | *Control of EVSE* |
| *5* | *Proximity detection* | *Proximity detection* | *Vehicle acknowledges plugged in* |

EVs and PHEVs come equipped with a Level 1 EVSE three-prong cord set so that no additional equipment is needed for charging—just plug in to a suitable rated electrical outlet such as NEMA 5–15 (120 V ac, 15 A). This is very slow charging at only a 2- to 5-mile range per hour of plug-in time. That is why Level 2 charging is preferred and EVs, depending on battery pack capacity, may come standard with a Level 2, 240 V ac, 3.3 kW, 6.6 kW OBC or higher. Figure 3.1 illustrates a sample of standard connectors, namely, the SAE J1772 Electric Vehicle and Plug-in Hybrid Electric Vehicle Conductive Charge Coupler.

Typical wall-plug-to-vehicle high-voltage (HV) battery efficiency depends on OBC rating (SAE level), the charging power level at which it operates, and the utility supply voltage. For example, a typical Level 2, 240 V ac, single-phase, 3.3 kW OBC operating near rated power is on the order of 92% efficient, and when connected to its EVSE at approximately 99%, the wall to HV battery efficiency is 90%. For Level 1 charging at 120 V ac, single-phase, 1.3 kW will be on the order of 90%, and its EVSE is approximately 98% efficient, for a system efficiency of approximately 88%. In both of these scenarios, the OBC is considered to be isolated with an active front end consisting of a boost-type power factor corrector (PFC).

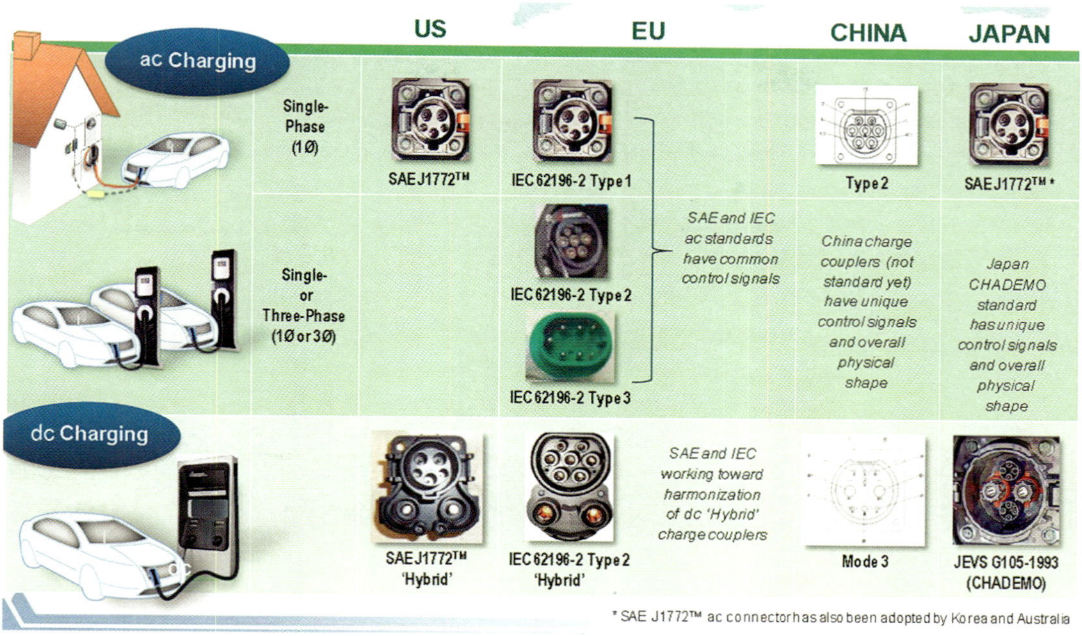


**Figure 3.1** Illustration of charge couplers.

The communications between the vehicle and EVSE must ensure, during the vehicle's plug-in period, that the vehicle drive mode is disabled, that proper charge indication such as instrument panel lights is given, and also that an indication is provided on the EVSE that charging is taking place. On the vehicle, a number of logic tests are executed before initiating charging. Figure 3.2 is provided to give some indication of what a conductive charging schematic may be for an EV having an optional dc fast-charge port. In this case, the vehicle battery is a nominal 360 V pack, and the OBC is Level 2 at 6.6 kW (240 V ac, 1-phase, 30 A). In this conductive charging system, a number of contactors, or relays, are employed to initiate certain control actions such as those to ensure that charge port activations are mutually exclusive and that battery galvanic isolation is maintained. Certain functions that receive high-voltage power during charging must be inhibited, such as the traction inverter, to prevent the vehicle from driving away while charging. Other functions are optional such as energizing the cabin heating ventilating and air conditioning (HVAC) via positive temperature coefficient (PTC) heating elements in the cabin climate control unit, or the air conditioning electric drive compressor. Control of HVAC functions are outside the scope of this chapter. However, a dc-dc converter to maintain low voltage systems is necessary, so the HV-to-14 V converter must operate during charging to power the electronic control modules (ECM) in the OBC and other units.

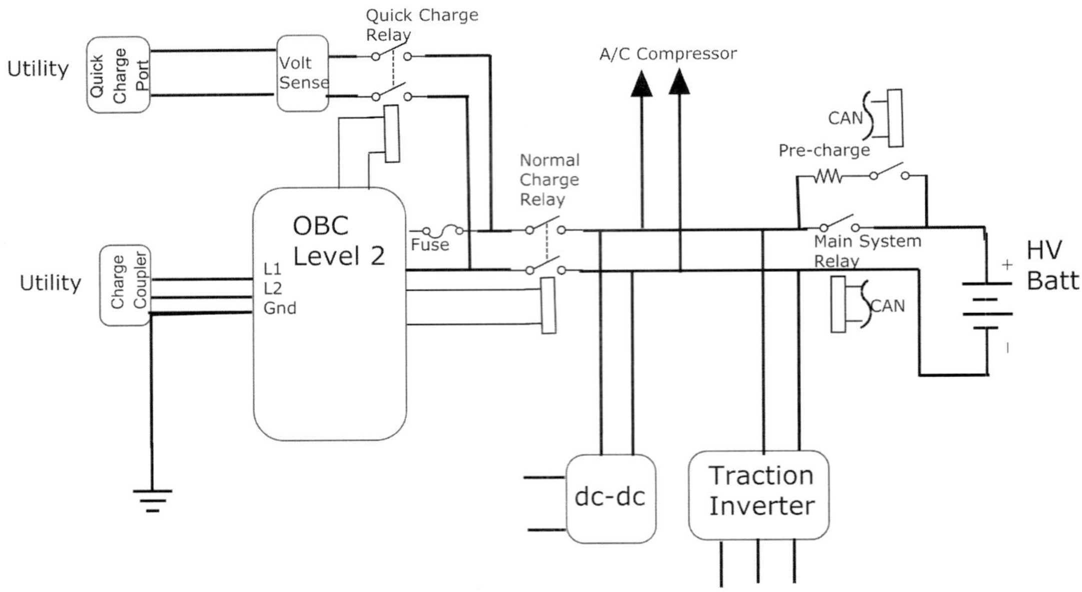


**Figure 3.2** Schematic of EV having Level 2 OBC and dc fast-charge port.

## 3.2 Wireless Charging

At the present time, wireless charging is an option that exists in addition to conductive OBC, as shown in Figure 3.2, and is most often implemented as an aftermarket product. As such, the proliferation of aftermarket WPT charging products is adding confusion to the marketplace because of the lack of standardization. Unlike the conductive charging standard, SAE J1772 cited in Section A, efforts towards the standardization of WPT by the SAE J2954 wireless charging task force are still at an early stage. The main objectives of WPT standardization are as follows:

- Interoperability of center frequency, vehicle secondary coil mounting location, radio communications, grid detection, and remote control
- Specification of misalignment tolerance, alignment aids, metallic and biological obstacle detection, and driver display
- Compliance with high-frequency magnetic and electric field emissions, interference with other services in adjacent and other frequency allocations, and emissions levels in public zones that include the passenger cabin and alongside the vehicle
- Power regulation control methods such as primary side only, or combination of primary and secondary side regulation, and autonomous (software) control
- Field testing and validation of new installations, and specification of standard testing hardware

Figure 3.3 is a functional diagram of WPT technology proposed by Oak Ridge National Laboratory that was developed for the U.S. Department of Energy [3-1]. Several key aspects of WPT technology are shown in Figure 3.3, including the following:

- The WPT charger operates exclusively or with existing conductive charging.
- Power regulation is primarily on the grid side, but secondary side rectification and enhancement to active regulation, such as a buck or boost for the power converter, are permitted.
- Communications occur via a bi-directional radio having access to the vehicle controller area network (CAN) as well as WPT specific signals as needed.
- Active front end; i.e., PFC, high-frequency isolation transformer in the primary pad cabling, and high-voltage contactor on the vehicle secondary side to isolate the WPT from the vehicle traction battery when not in use.

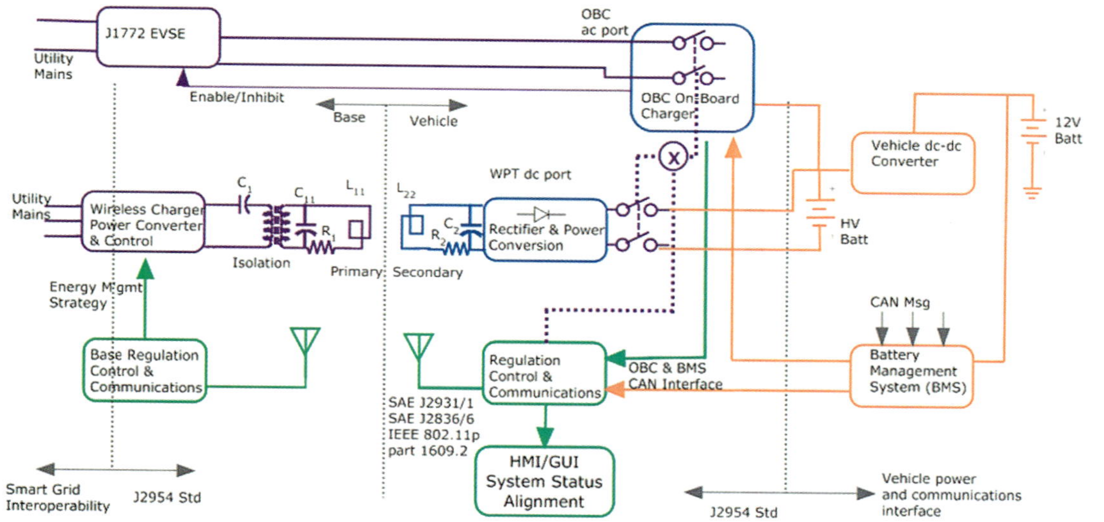


**Figure 3.3** Functional diagram of wireless charging option to OBC.

A provision in WPT provides the vehicle operator with information on proper alignment to a primary charging pad and displays this in some appropriate manner in the vehicle passenger cabin human machine interface (HMI) along with necessary system status and fault signaling. For example, on approach to a charging location, the V2I radio will request some level of excitation of the primary charging pad or ancillary services to aid in proper approach and alignment; this information is then displayed to the driver. In addition, battery state-of-charge (SOC) status may be displayed and, from this, an appropriate message to the WPT base unit on an initial power level request for charging and potentially some interaction with grid services for time of use (TOU), utility rates, and load shedding in the case of a utility operating near capacity. Many other options for information exchange and display are possible.

Just as in conductive charging, the efficiency of power transfer in WPT is crucial, with guidelines and most implementations calling for parity between the two methods. This means that WPT should provide > 90% utility to battery charging efficiency under conditions of compatible WPT hardware, coupling coil gap, and alignment. In cases of less than optimum compatibility of WPT hardware and alignment, efficiency is then relaxed somewhat to > 85% utility to the vehicle battery. However, debate continues as to whether this efficiency metric includes charging power diversion via the vehicle dc-dc converter to 12 ancillaries (necessary during any type of charging) and other losses. In addition, the need is pressing to identify which power level, or range of power and other conditions, the efficiency metrics must meet. Figure 3.4 shows the WPT functional cascade starting from the 240 V ac utility connection, active front end (AFE) containing PFC electronics, a high-frequency (HF) isolation transformer for electrical safety, the WPT coupler, secondary side rectifier plus filter, and the vehicle HV battery.

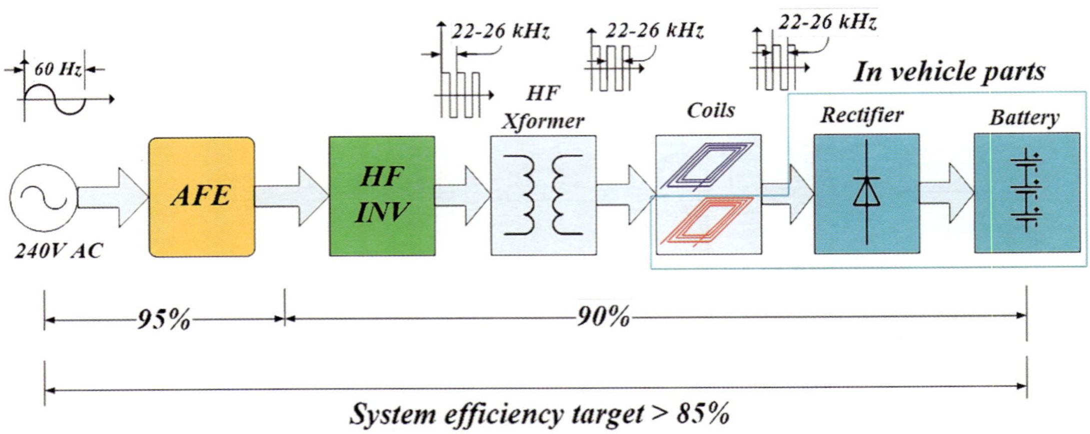


**Figure 3.4** Wireless charging functional cascade and representative signal types.

Considerable similarity exists between conductive and wireless charging functionality. This can be appreciated in reference to Figure 3.4 and in comparing to the functional requirements of an OBC as discussed. Because both conductive and wireless charging must meet utility power quality specifications, both require an AFE. The AFE contains a nominal power factor corrector stage, essentially an active rectifier, which forces the utility ac line current to be in phase with the line voltage to maintain a power factor (PF) of > 0.95. Both the OBC and the WPT base units rely on HF ac to implement their required functions, including galvanic isolation of the dc charging from the grid ac line. In the OBC, this is accomplished with a high-frequency transformer as part of the regulation electronics, and in the WPT, this is accomplished by adding a high-frequency isolation transformer. Even though the primary coil to secondary coil provides isolation of the vehicle from the grid side, it does not provide isolation of the primary charging pad from the grid. The inclusion of a HF transformer achieves galvanic isolation in addition to providing the WPT designer with added flexibility in design of the HF power inverter stage, such as rail voltage specification, power semiconductor rating, and capacitor ratings, without affecting the design of the primary coil circuit. Operational principles of WPT will be given in detail in subsequent chapters.

A HF transformer design example is provided here to acquaint the reader with coupling coefficient, k, by contrasting a tightly coupled 10 kVA ferrite cored coaxial winding transformer (CWT) with that of a similarly rated set of coupling coils. It is well known that commercially available soft ferrites, such as manganese-zinc (MnZn), have relatively low induction (Bs <0.38 T), certainly in comparison to motor grade electrical steel (Bs>1.8 T). Because of the high frequency, it is possible to obtain a very large reduction in overall size. This can be seen by referring to the 10 kVA CWT shown in Figure 3.5 and noting its construction details provided in Table 3.3. This CWT has been designed with a 3.75:1 turn ratio so that the relatively low source point voltage of the series tuned WPT primary can be matched to a power inverter operating with a dc link voltage of 400 V dc to 720 V dc or higher. The higher rail voltage of the HF power inverter means lower switching current and, hence, higher efficiency.

**Figure 3.5** HF coaxial winding transformer rated 10 kVA.

**Table 3.3 Core and Winding parameters for the 10 kVA CWT**

| Core Parameter | Unit | Value | Winding Design | Unit | Value |
|---|---|---|---|---|---|
| Outside diameter, OD<br>Each stack | (mm) | 88 | Primary turns, Np | (#) | 15 |
| Inside diameter, ID<br>Each stack | (mm) | 64 | Secondary turns, Ns | (#) | 4 |
| Stack, L | (mm) | 98 | Turns ratio | (a) | 3.75 |
| Core mass (3C90)<br>Both stacks | (kg) | 3.28 | Magnetizing current, Im @ Us=556Vrms, 23 kHz | (Arms) | 2.6 |
| Core loss factor, Pv@25 kHz | (W/lit) | 100 | Primary current Ipri<br>@ 10 kW | (Arms) | 32 |
| Core saturation flux density, Bs | (T) | 0.38 | Secondary current Isec @ 10 kW | (Arms) | 120 |
| Coupling coefficient, k | (#) | 0.99 | Primary current density, Jpri @ 10 kW, #6 AWG Litz | ($A/mm^2$) | 2.6 |
| Primary leakage inductance, Llp | (μH) | 84 | Secondary current density, Jsec @ 10 kW, #2 AWG Litz, | ($A/mm^2$) | 3.57 |
| Secondary leakage inductance, Lls | (μH) | 80 | Total CWT mass, Mcwt | (kg) | |

To complete the comparison, Figure 3.6 shows the coupling coil pair necessary for a transfer of >10 kVA power across a gap of 160 mm and up to 20 kW peak power. These coils have an outside diameter of 720 mm over the #2 AWG Litz cable showing in the lower coil. The primary and secondary coils are wound with eight turns each and have an inductance of approximately 84 $\mu$H and a coupling coefficient at z = 160 mm of k = 0.24. The secondary coil is wound with #6 AWG Litz cable because it will not carry magnetizing current to the extent that the primary will. The coupling coil pair, if set one on top of the other, would occupy a volume of 787 × 787 × 100 mm, whereas the CWT occupies a total volume of approximately 176 × 176 × 88 mm, including the end turns. The size difference is dramatic, and both use soft ferrite flux guides. This comparison illustrates the challenge of WPT in the case of large separation between the coils.

**Figure 3.6** Coupling coils rated 10 kVA continuous, dimensions: 787 × 787 × 50 mm.

## 3.3 Commercial WPT Technologies

The following short summaries are given to provide insights into some of the international programs now underway, or about to start, that are focused on finding the benefits of wireless stationary and dynamic (i.e., in-motion) charging.

The London Trial [3-2] is designed to promote EVs through convenient WPT charging and the benefits it entails. This activity was organized in 2012 by the UK Office for Low Emission Vehicles, Transport for London, and partners Qualcomm Halo IPT, Charge Master, Renault, delta motorsport, and Addison Lee. Qualcomm and Renault plan on WPT-equipped vehicles in operation in fleet and private use by mid-2013, followed by installation of a taxi fleet.

Another WPT trial is the Berlin INTEROP [3-2] project, which is designed to prove the technical and economic feasibility in the public sector through the installation of 70 charging spaces. The trial includes a fleet of some 50 vehicles from different manufacturers equipped with WPT units from four different suppliers. Like the London Trial, it also started in 2012 and is planned to run through 2015. The WPT suppliers include Conductix-Wampfler, Vahle Electrification Systems, SEW Eurodrive, and Siemens.

The preceding WPT evaluation trials are focused on stationary wireless charging. Another initiative, European Project FABRIC (Feasibility Analysis and development of on-Road charging solutions for future electric vehicles) [3-2], was commissioned under the European Green Cars Initiative, Seventh Framework Program, approved by EC FP7 from the year 2014 to 2016, and coordinated by Fiat and 29 partners from several EU nations. The demonstration program calls for roadway integration of primary coil tracks energized from an energy management roadside unit and wireless LAN for V2I communications that include charging requests, regulated speed and lane departure warnings, and messages for authorization, traffic control, and billing.

In addition to the suppliers just mentioned, other vendors that also provide WPT charging for light-duty vehicles include Halo-IPT, SEW-Brose, WiTricity, and KAIST. When the list is expanded to include research organizations, it expands very rapidly with the ORNL, the Energy Dynamics Laboratory (EDL) at Utah State University, and many others. Each of these suppliers offers preferred coupling coil designs, operating frequencies, and power regulation strategies. The reader is referred to the SAE wireless charging task force, J2954, for more details [3-3].

## 3.4 Rollout of Wireless Charging

Wireless charging relies on high-frequency magnetic fields to couple power from a primary coil to a secondary coil that may be separated from 125 mm to 200 mm, typical of passenger vehicle ground clearance. Safety, both electrical and from the coupling fields, must be given the highest priority. For example, as shown in Figure 3.7, a person standing near the vehicle under charge must not be exposed to high-frequency magnetic or electric fields exceeding international specifications such as ICNIRP2010 (International Commission on Non-Ionizing Radiation Protection), which specifies that whole body exposure should be less than 27.3 $\mu$T (micro-Tesla). For WPT, the frequency band from 10 kHz to 148 kHz is of most interest, and most WPT systems in operation fall within this range.

(a) Specified emissions at 1.3 m from the primary coil center (courtesy ORNL).

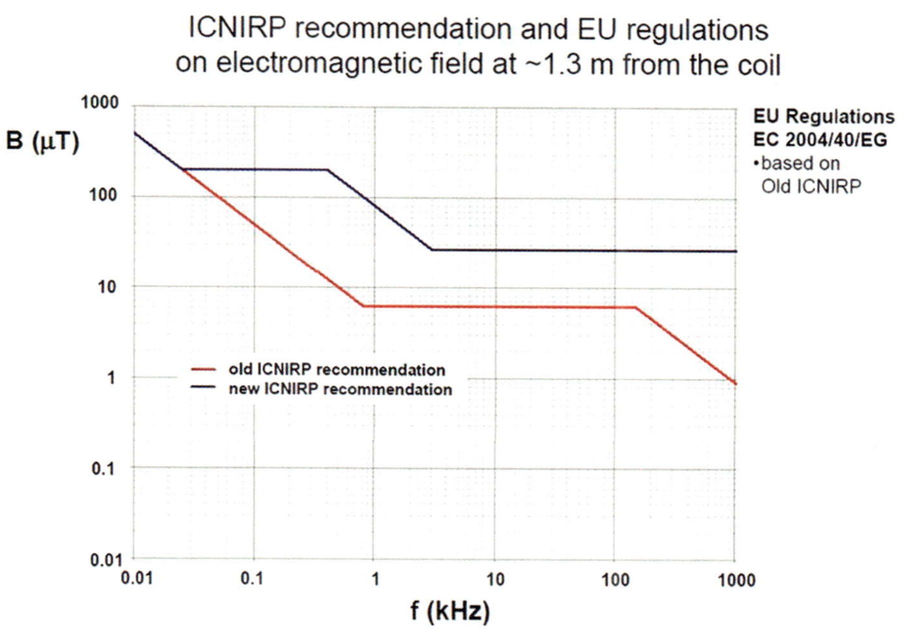


(b) ICNIRP2010 specification.

**Figure 3.7** Magnetic and electric field emissions for safety.

As WPT technology matures, it will become more pervasive in light-duty (LD) vehicles for stationary charging, as shown in Figure 3.8 for a PHEV. Typical charging power levels for LD vehicles will fall into SAE Level 2, which spans 3.3 kW to 19.2 kW levels. For example, a Chevy Volt extended range electric vehicle (EREV) will charge at 3.3 kW maximum to its nominal 330 V 16 kWh lithium-ion battery pack. A Nissan LEAF®, a BEV will charge at 6.6 kW nominal via its production OBC, or at 45 kW dc fast charge via an optional fast charge port, to its nominal 365 V 24 kWh lithium-ion battery pack. The Toyota Motor Company RAV4 is designed to charge at 10 kW via a production OBC to its 386 V nominal 41.8 kWh lithium-ion pack. As another example, the GM Spark EV to be introduced in 2014 will charge at 6.6 kW to its 406 V nominal 20 kWh lithium-ion battery pack.

WPT essentials consist of grid connection of appropriate capacity, a vehicle electric ESS of commensurate rate capability, and two-way communications to manage the charge process.

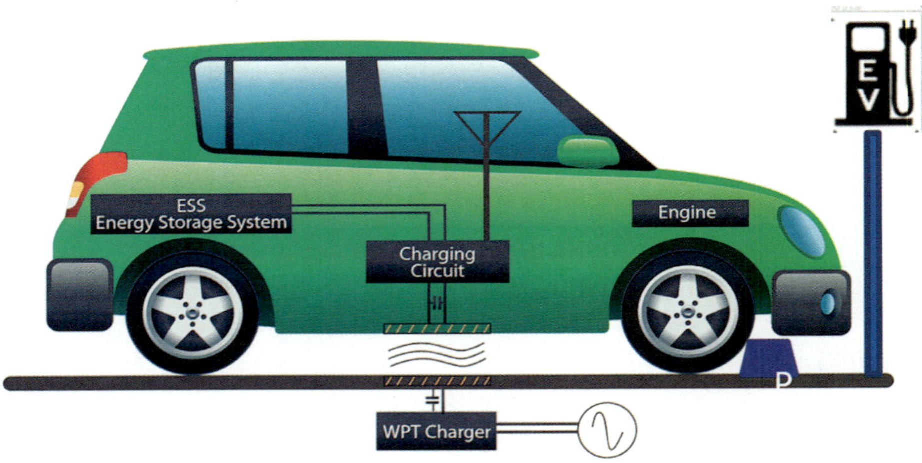


**Figure 3.8** PHEV and wireless charging equipment (courtesy ORNL).

As WPT charging matures beyond SAE Level 2, it will be economical on heavy-duty (HD) vehicles such as shuttle and transit buses. For example, a WPT charger operating at dc fast charge levels of 45 kW and higher would be needed to opportunity charge a transit bus when it makes scheduled stops along a fixed route, as shown in Figure 3.9. A transit bus, for example, may require charging at levels above 150 kW, depending on the ESS rate capability, SOC, temperature, and equipment ratings.

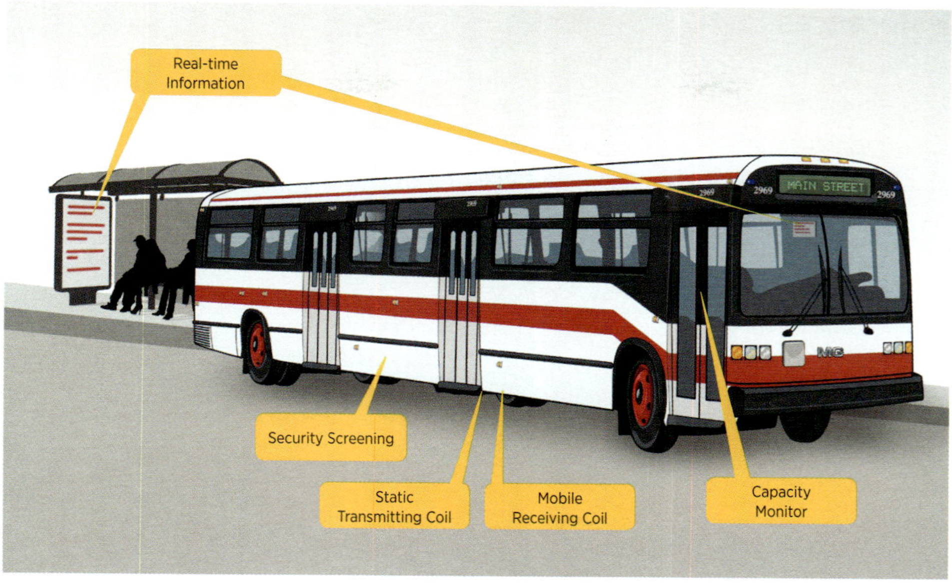


**Figure 3.9** Wireless opportunity charging HD vehicles (courtesy ORNL).

When WPT charge capability transitions into the dc fast charge to high-rate HD vehicle charging, it will be well positioned to take on dynamic charging, which is inherently high power and short duration. During in-motion charging, and assuming that 100% energized highways are not economically viable, then energized tracks of appropriate length and with optimal repetition patterns will be feasible. In-motion WPT is a concept that will have the following benefits to all electric vehicles (AEVs):

- Unlimited range because power comes from the roadway
- Potential to downsize onboard battery packs and thereby reduce vehicle mass
- Facilitation of eventual fully autonomous travel by accelerating development of energized guide ways

However, all this does not come without issues. Energized roadways may need to be dedicated lanes restricted to LD vehicles so that embedded coils are not subjected to HD vehicle traffic. Furthermore, alignment during in-motion charging may require more dynamic lane following via higher-performance longitudinal stability programs on the vehicle. Electrical burden on the utility will be more intermittent and time-of-day-dependent. And finally, communications such as V2I will be required to have very low latency so that coil energizing is properly sequenced to follow each vehicle.

These topics and more will be discussed in depth in the next chapter.

## References

3-1. J. M. Miller, O. C. Onar, D. E. Smith, P. T. Jones, "ORNL's Wireless Charging Program," presentation to SAE J2954 committee, 27 March 2013.

3-2. J. Perrin, "Inductive Charging of Electric Vehicles: A European Perspective," 2nd Conference on Electric Roads and Vehicles, CERV2013, Park City, Utah, 4–5 February 2013.

3-3. SAE International, www.sae.org, SAE J2954, chair: Jesse Schneider, Schneider.jesse@web.de.

# Chapter 4
# An Overview of OLEV Technology

## 4.1 Background

The serious environmental pollution caused by internal combustion engines together with the depletion of fossil fuels has motivated global interest in eco-friendly energy [4-1], [4-2], [4-3]. Notably, electric vehicle technology has been developed to reduce the use of fossil fuels in vehicles, which are the main consumers of fossil fuel [4-4], [4-5] [4-6], [4-7]. As a result, hybrid electric vehicles that use both a combustion engine and an electric engine have already been widely commercialized [4-4]. However, all-electric vehicles, such as plug-in electric vehicles and battery electric vehicles, are in marginal market growth at present owing to battery-related drawbacks such as large size, heavy weight, high price, long charging time, and short driving range. These problems cannot be solved easily with the current battery technology. In an effort to address battery problems, the concept of roadway-powered electric vehicles has been proposed [4-8], [4-9]. With this system, the electric vehicle is charged on the road wirelessly, and the battery can, hence, be downsized; additionally, no waiting time for charging is needed. Much research on wireless power transfer for electric vehicles has been performed over the past few decades. The Partner for Advanced Transit and Highways (PATH) project of UC Berkeley demonstrated a roadway-powered electric vehicle system capable of transferring power with a 60% efficiency across an 8 cm air gap [4-7]. In this project, a powered roadway track was constructed and was experimentally validated. Loosely coupled inductive power transfer systems have been proposed to overcome the large air gap for practical operation on roadways. To achieve high power transfer efficiency, many techniques, including resonant inverters for wireless power transfer, efficient pick-up modules, effective pick-up tuning methods, and pick-up voltage control methods have been proposed.

The OLEV center of the Korea Advanced Institute of Science and Technology (KAIST) has developed a high-efficiency roadway-powered electric vehicle system [4-10]. The

OLEV system achieved 100 kW output power with 80% power transfer efficiency of the system across a 20 cm air gap [4-11], [4-12]. The first generation of the OLEV system has already been commercially implemented in Seoul Grand Park [4-13], [4-14], and the second generation of the OLEV system has been commercialized at KAIST campus and Gumi city in Korea. Figure 4.1 shows earlier versions of prototypes of the OLEV vehicle as of the year 2009 in (a); the historic first public launch of the OLEV train in March 2010 in (b); the currently daily operating shuttle bus on the KAIST campus in (c); and the first public transit OLEV bus of Gumi City in (d).

(a) (b) (c) (d)

**Figure 4.1** OLEV vehicles: (a) Earlier prototypes of various OLEV vehicles in 2009, (b) The first public launch of the OLEV train in March 2010, (c) Campus shuttle bus in KAIST, (d) The first public transit OLEV bus of Gumi City.

## 4.2 SMFIR Technology

With different lay-ups of power cable combined with different geometries of the ferrite core at the bottom of the cable, the magnetic field shapes, paired with the design of the pick-up devices, and the resultant efficiency of wireless power transmission can vary [4-7], [4-12]. By placing the ferrite structures in an optimized way, the magnetic field shape can be structured so that high-density magnetic flux can be transferred to the pick-up coils [4-12], as shown in Figure 4.2. Numerous design iterations have been performed to achieve the maximized power transmission to the vehicle while reducing the magnetic field intensity to the allowable leakage field around the vehicle. Figure 4.3 shows an example of various magnetic field shapes during the simulations.

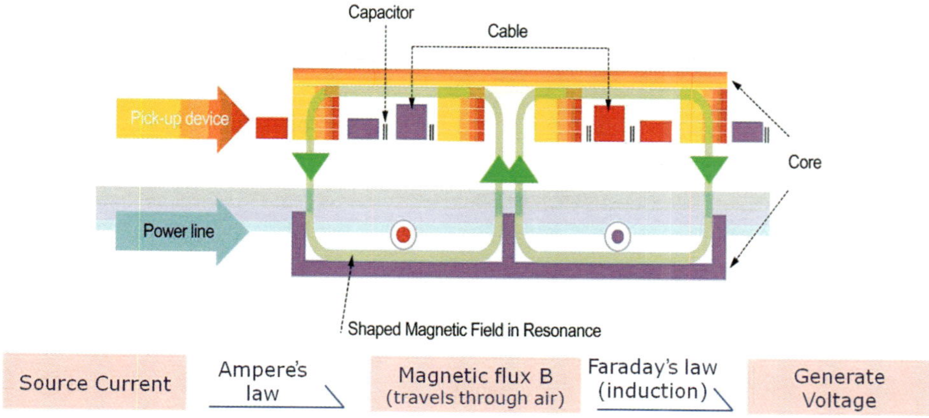


**Figure 4.2** OLEV power transmission based on shaped magnetic field in resonance (SMFIR).

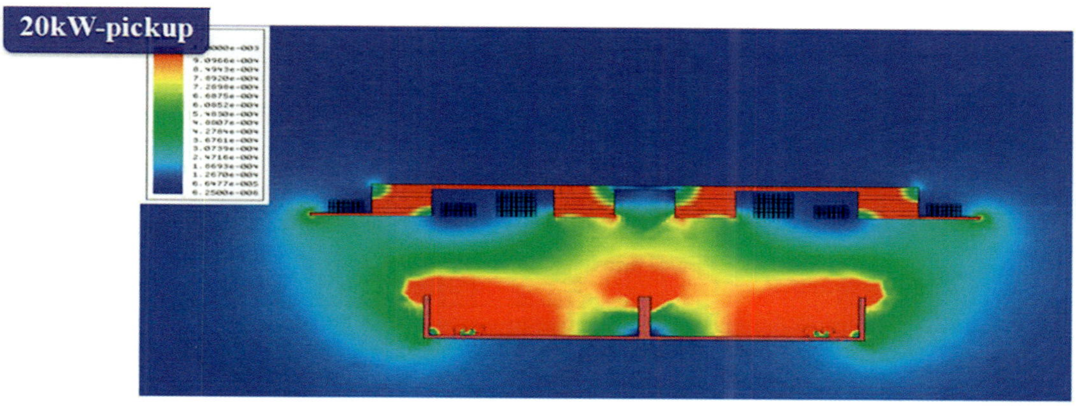


**Figure 4.3** Simulation result of SMFIR at 20 kHz.

With the SMFIR technology, the OLEV system has been proven to be able to transfer 100 kW of power across a 20 cm air gap with more than 80% power transfer. The permissible EMF exposure to human users has been kept below 62.5 mG, following the ICNIRP guideline of 1998 [4-15].

## 4.3 Overall System

### 4.3.1 System Operation

A wireless power transfer system uses magnetic resonance coupling [4-16], [4-17], [4-18], [4-19], [4-20], [4-21], [4-22], [4-23], [4-24]. One of the most important factors that must be considered in designing a magnetic resonance system is the target power of the system.

The voltage and current ranges of usable devices and operating frequency of the system will depend on the required power of the target application [4-10], [4-12].

Because the wireless power transfer system for moving electric vehicles is a public service system that is installed on a road, the use of the resonance frequency must be permitted by the applicable regulations. Generally, wireless power transfer systems for electric vehicles use frequencies in the range of 10–100 kHz. In the OLEV system, the target power is 100 kW and the resonance frequency is 20 kHz [4-6], [4-7], [4-10], [4-12].

Figure 4.4 is a conceptual schematic diagram of an OLEV bus system. A wireless power transfer system consists of a power transmitter and a power receiver. The power transmitter is composed of an inverter and power lines. The inverter provides power, and the power lines carry current and generate magnetic flux. The power receiver consists of pick-up modules, rectifiers, and regulators. The pick-up modules generate power from induced voltage and current, the rectifiers convert ac power to dc, and the regulators control the output voltage, which is input to batteries and motors.

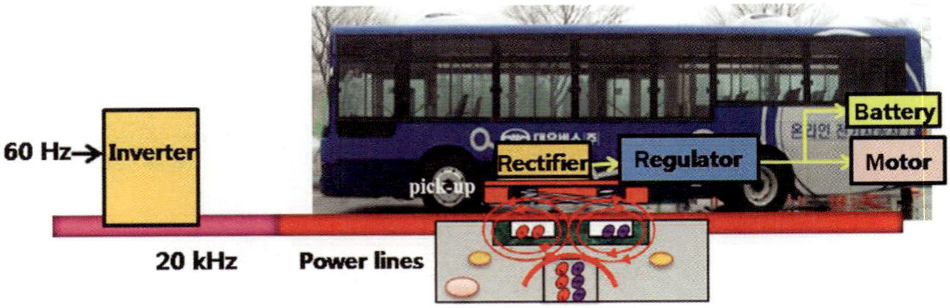


**Figure 4.4** Conceptual diagram of the OLEV system (© 2014 IEEE. Reprinted with permission, from [4-12]).

The inverter receives power from an electric grid and converts the typical 60 Hz of the grid operating frequency into a 20 kHz resonance frequency of electricity. Although the inverter can be controlled to provide constant voltage, constant current control is more advantageous in dealing with changes in the load resistance or multi pick-up charging. Therefore, in the OLEV system, the inverter converts 60 Hz power to 260 A constant current at 20 kHz resonance frequency. The power line modules are installed underneath the road and along the road. For the construction and operation of the power supply system on the road, the location of the power line can be at the start point, destination point, and stopping area, where the vehicles stay longer on the power line during a longer time of charging for better use of the power line. Another location will be the uphill section, which requires additional driving power from the motor, so that the wirelessly delivered power can be additional power directly to the motor added to the onboard battery power [1-25], [1-26]. This feature will help to reduce the energy storage capacity of the onboard battery, which will be one element for business analysis of the OLEV system. Power line cables in the module generate magnetic flux, and the magnetic cores make the flux propagate upward. In the pick-up modules, the magnetic cores capture the magnetic flux from power line modules, which induces voltage along the coils. Compensation capacitors are connected to the coils to compensate for the impedance of the inductance. The regulators are dc-dc converters, which control the effective load resistance to control the voltage and current of the pick-up modules. A buck converter, a boost converter, or a buck-boost converter may be used to increase or decrease the voltage. In the OLEV system, a boost converter is used to consider the battery input voltage.

Some of the transferred power is used to drive the motors, and the remainder is used to charge the batteries. When the vehicle stops, all of the power is used to charge the batteries.

### 4.3.2 Segment Operation

The power lines of the OLEV system are divided into several power line segments, as shown in Figure 4.5. For operational efficiency concerns, the inverter turns on only those segments on which a vehicle is located. A sensor in a segment senses the approach of a vehicle, sends a message to the inverter, and the inverter turns the segment on. If the vehicle moves to the next segment, the inverter turns the previous segment off and the next segment on. This segment operation method reduces power loss in the cables and also reduces EMF exposure to people on and near the powered road.

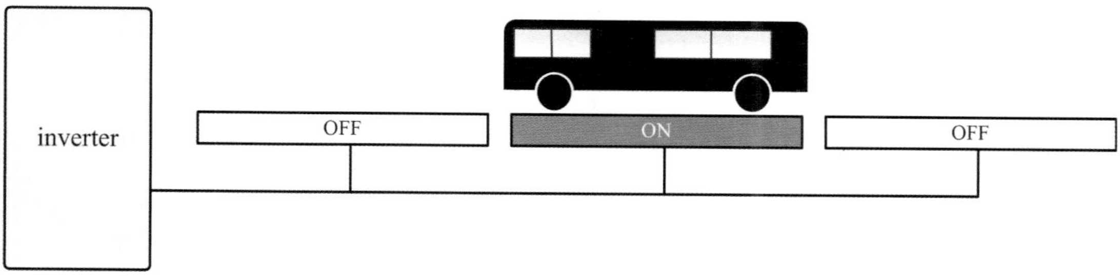


**Figure 4.5** Concept of power line segmentation method (© 2014 IEEE. Reprinted with permission, from [4-12]).

## 4.4 Design of Power Supply Infrastructure

### 4.4.1 Road-Embedded Power Systems

The power supply systems for the proposed system are shown in Figure 4.6. The road-embedded power circuit consists of two parts: a three-phase power inverter and a road-embedded rail. The power inverter converts three-phase 60 Hz ac voltage to a constant, single-phase 20 kHz ac current. In the internal circuit of the power inverter, three-phase ac voltage is rectified to dc. This dc voltage is converted to an isolated single-phase ac voltage source by a single-phase inverter. The inductance of the power cable in the embedded road rail $L_1$ has a value that varies with the length of the coil.

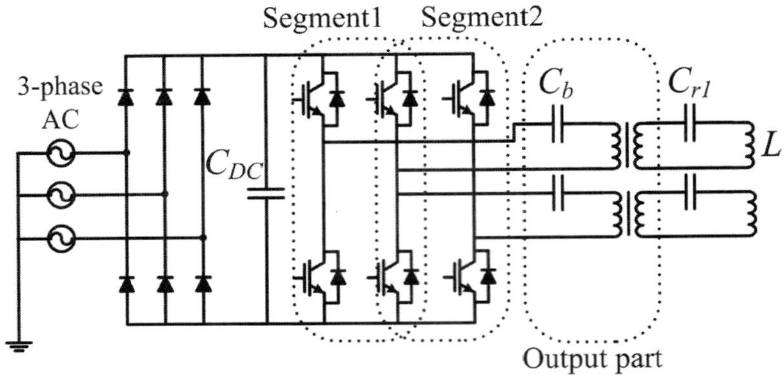


**Figure 4.6** Power circuits for proposed inverter system (© 2014 IEEE. Reprinted with permission, from [4-12]).

The coil is divided into two segments. Meanwhile, based on $L_1$, the resonant capacitance $C_{r1}$ is adjusted to meet the resonant frequency of 19.8 kHz, which is slightly below the inverter switching frequency. The equivalent impedance of the embedded coil including $C_{r1}$ operates as a small inductance at a no-load state, and its current has a 90° lagging phase compared to the inverter output voltage. Its equivalent impedance becomes almost a resistor at full-road state, and its current has the same phase as the inverter voltage.

### 4.4.2 Inverter and Electrical Segmentation Design

The power line current is changed according to various load conditions: for example, the drive pattern of a car or the state of charge (SOC) of batteries. However, supply of constant current is necessary, regardless of whether the transfer power level is high or not, because we want a uniform magnetic flux density to support steady induction voltages in the pick-up modules. Therefore, an output current regulator is designed using a pulse-width modulated (PWM) inverter. In Figure 4.7, the controller consists of a proportional-integral (PI) control loop. The PWM method is used for steady-state switching of a full-bridge inverter for one segment.

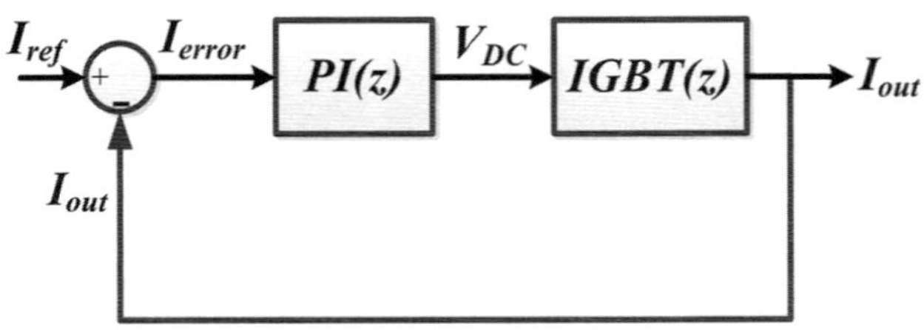


**Figure 4.7** Output current controller of proposed inverter system (© 2014 IEEE. Reprinted with permission, from [4-12]).

Segmentation is accomplished using an electrical switching scheme, as shown in Figure 4.6. Without using mechanical switches, which have poor performance and a short life, we can switch from one segment to another using a different insulated-gate bipolar transistor (IGBT) arm.

### 4.4.3 Power Line Module

The power line cables are located on the primary ferrite core, and 260 A/20 kHz currents are supplied from the inverter. These currents consistently generate magnetic flux for the pick-up. A Litz wire structure is used for the power line cables to avoid the skin effect. The cross-sectional area of the power line cables is nearly $95\text{mm}^2$. The area issue is important when a high current is supplied to the cable. Our analysis result proves that an area of $1\text{mm}^2$ can endure current of 3 A.

Fiber-reinforced plastic (FRP) tubes are used to prevent the power cables from being broken. FRP is durable enough to tolerate the weight of vehicles and is electrically and magnetically transparent. The FRP tubes physically protect the power line cables from underground pressure, heat, and water.

To operate the power line segmentation method, magnetic sensors are located at the beginning of the segment to notice the arrival of electric vehicles. When an electric vehicle is sensed by the sensor of the first segment, the first power line segment turns on. And when the second sensor senses an electric vehicle, the first power line segment turns off and the second segment turns on. The signal cable delivers the notification message from the sensor of the power line module to the inverter controller.

The power line module consists of ferrite cores, segment power cables, signal cable, common power cables, FRP tubes, and aluminum tubes, as shown in Figure 4.8. The FRP tubes are used to protect the power cables and sensor signal cables. Common power cables are used to connect each segment to the inverter and need to be shielded so they do not affect segment power cables magnetically. Thus, aluminum tubes are used for the common power cables. Aluminum has low permeability and high conductivity, which are good shielding characteristics against magnetic flux.

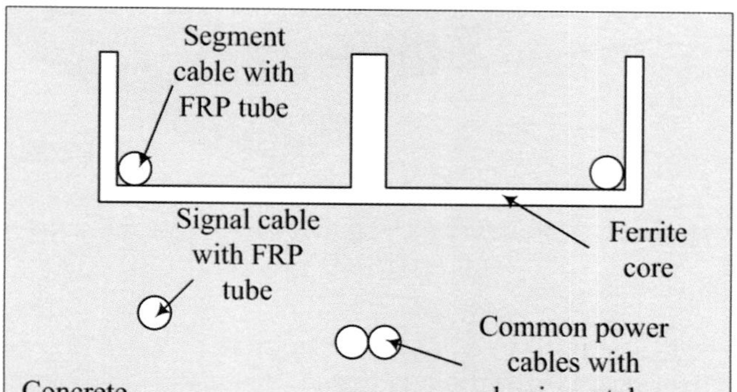


**Figure 4.8** Structure of proposed power line module (© 2014 IEEE. Reprinted with permission, from [4-12]).

## 4.5 Design of Power Collection Systems

### 4.5.1 Pick-Up Module

The pick-up module is composed of ferrite core blocks, pick-up cables, compensation capacitors, and an FRP case. The case protects the inside physically and insulates the outside from electrocution. Similar to the power line cables, pick-up cables are Litz wires. Figure 4.9 shows a pick-up module without the cover. The pick-up cables are wound around the ferrite cores. In the OLEV system, the center winding has 64 turns of pick-up cables, and the left and right windings have 28 turns, respectively. To add up the induced voltages, the center winding is clockwise and the other windings are counter-clockwise. One half of the center winding is connected to the left winding, and the other half is connected to the right winding. Capacitors are connected to the cables to compensate for the inductance of the winding cables. To make the impedance zero at the operating frequency, the capacitance is selected using Equation 4.1

$$C_2 = \frac{1}{\omega_0^2 (L_2 + M)} = \frac{1}{(2\pi f_0)^2 (L_2 + M)} \tag{4-1}$$

*where* $C_2$ is the tuned capacitor of the secondary coil, $L_2$ is the inductance of the secondary coil, at the target resonance frequency of $f_0$ and the mutual inductance of $M$.

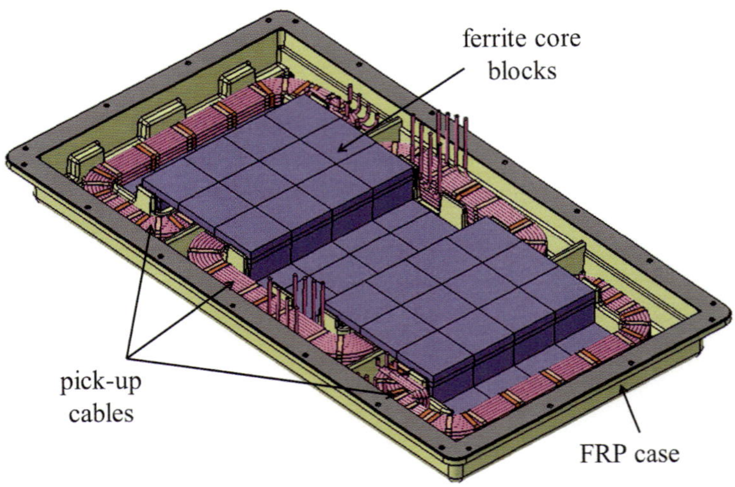


**Figure 4.9** Structure of proposed pick-up module (© 2014 IEEE. Reprinted with permission, from [4-12]).

In the OLEV pick-up, the reactive voltage due to a very large L is near 10 kV. This high voltage is difficult to insulate even if compensated. Hence, the capacitance is not connected all at one point but divided and distributed among the winding cables to reduce the peak voltage. This connection method drastically reduces the peak voltage caused by the inductances.

### 4.5.2 Rectifier and Regulator

The rectifiers convert ac current to dc current. The regulator boosts the voltages of the rectifiers up to a reference voltage suitable for battery charging. Five 20 kW pick-up modules are used to obtain 100 kW power capacity in the OLEV system, and each pick-up module has two separated windings. Therefore, the OLEV system has ten rectifiers and one regulator for ten windings. As shown in Figure 4.10, the inputs of ten boost converters are connected to the outputs of rectifiers, respectively, and the outputs of ten boost converters are connected to a battery. The ten boost converters' output voltages are controlled by adjusting the ten PWM signals of the IGBTs. The OLEV regulator controls the ten phases of the PWM signals so that each difference among ten phases uniformly becomes 36°. As a result, output voltage ripples are minimized by minimizing the sum of the input current variations. Figure 4.11 represents the input currents of ten boost converter modules when the duty ratio is 0.5. Figure 4.12 shows output voltage ripples of ten boost converter modules according to the number of IGBT PWM signal phases. As the number of phases becomes larger, the output voltage ripples of the regulator become smaller under duty cycle control.

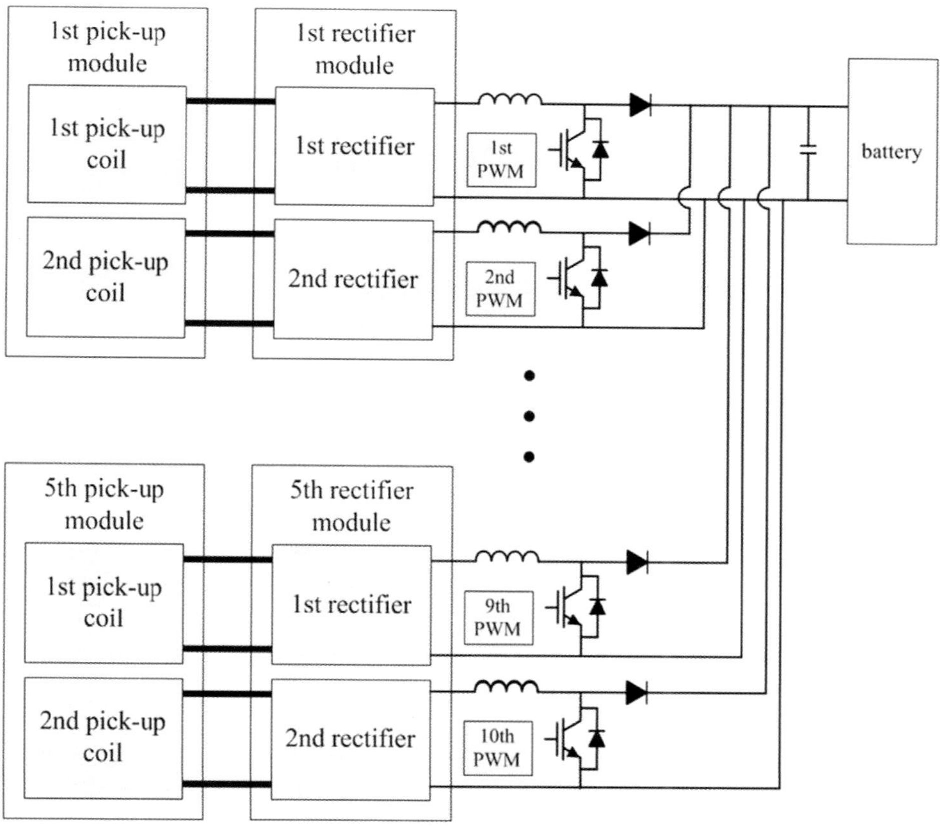


**Figure 4.10** Functional diagram of the OLEV power receiver system (© 2014 IEEE. Reprinted with permission, from [4-12]).

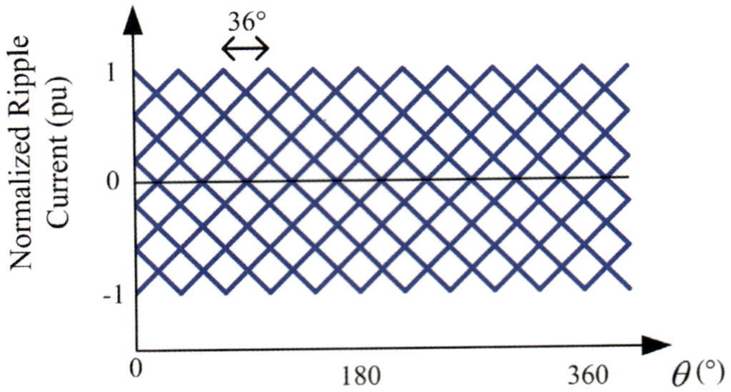


**Figure 4.11** Input ripple currents of ten boost converters (© 2014 IEEE. Reprinted with permission, from [4-12]).

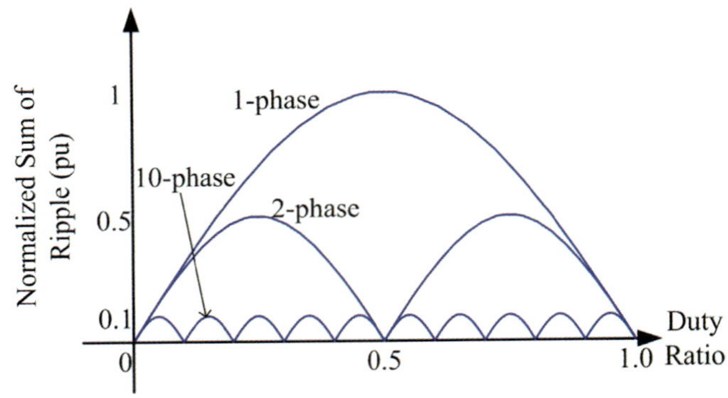


**Figure 4.12** Output voltage ripples of regulator for the numbers of phases of boost converters (© 2014 IEEE. Reprinted with permission, from [4-12]).

## 4.6 Application of SMFIR Technology to Bus

### 4.6.1 OLEV Power Supply Infrastructure

The OLEV power supply system or powered road is composed of a series of powered tracks. A powered track is composed of a series of segments, which can have different lengths from each other, controlled by a set of power inverters through a switch box. Figure 4.13 shows one example of a powered track that consists of six segments, where the first segment has a 3 m length, while the others have a 25 m length each. There can be sufficient distances between the segments depending on the surrounding conditions, and considering the cross-walk or bridges, etc. In Figure 4.14, we can see a realistic example of one powered track installation, which again shows that the powered track consists of six segments whose power engagement is controlled through a switch box of the inverter. The power supply control to each segment from the inverter, through the switch box, is triggered by the vehicle identification sensor installed under the road of each segment.

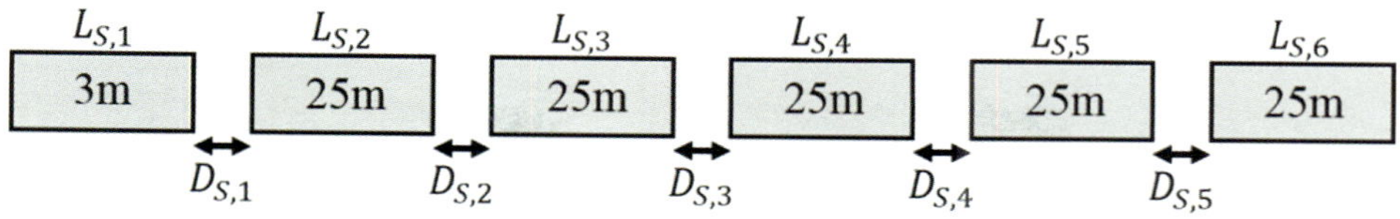


**Figure 4.13** An example of a powered track with six segments.

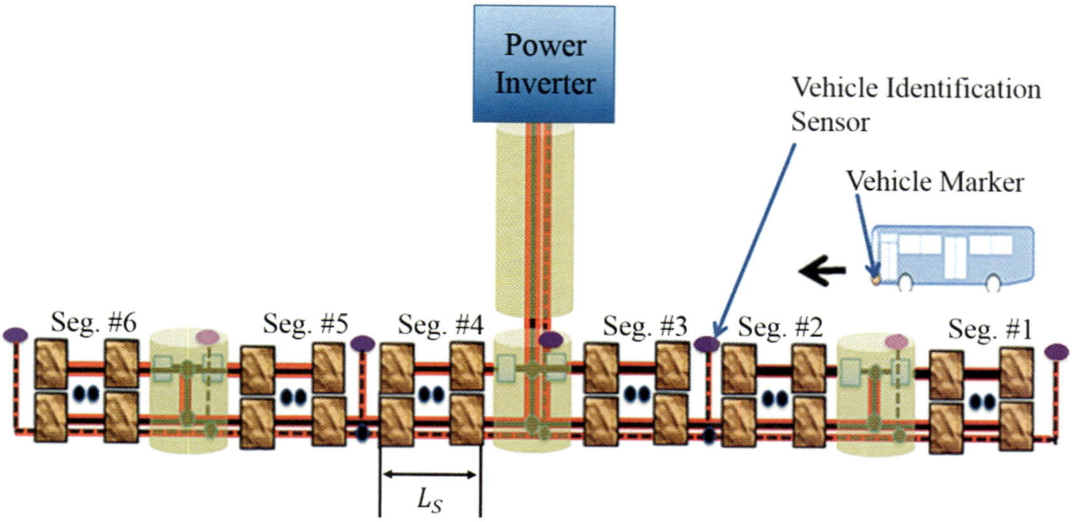


**Figure 4.14** A realistic example of one powered track.

One example of a powered road in Seoul Grand Park is shown in Figure 4.15. One round trip in a closed circuit of the people mover train is 2.2 km, and the total installed powered track is about 372.5 m, which is only 16% of one round trip. This system has three powered tracks and two supplementary stationary charging stations on the road. Three power inverters are installed at each powered track, while one power inverter is installed for two supplementary stationary charging stations. The electricity collected during vehicle operation, such as driving or stopping, on the 372.5 m powered track has been enough to compensate the battery SOC level, which is consumed during one round trip due to the drive motor and auxiliary power requirements of the OLEV train shown in Figure 4.1 (b). In other words, it is necessary to determine the powered road length considering the energy consumption of the vehicle on the application field while minimizing the installation cost [4-25], [4-26].

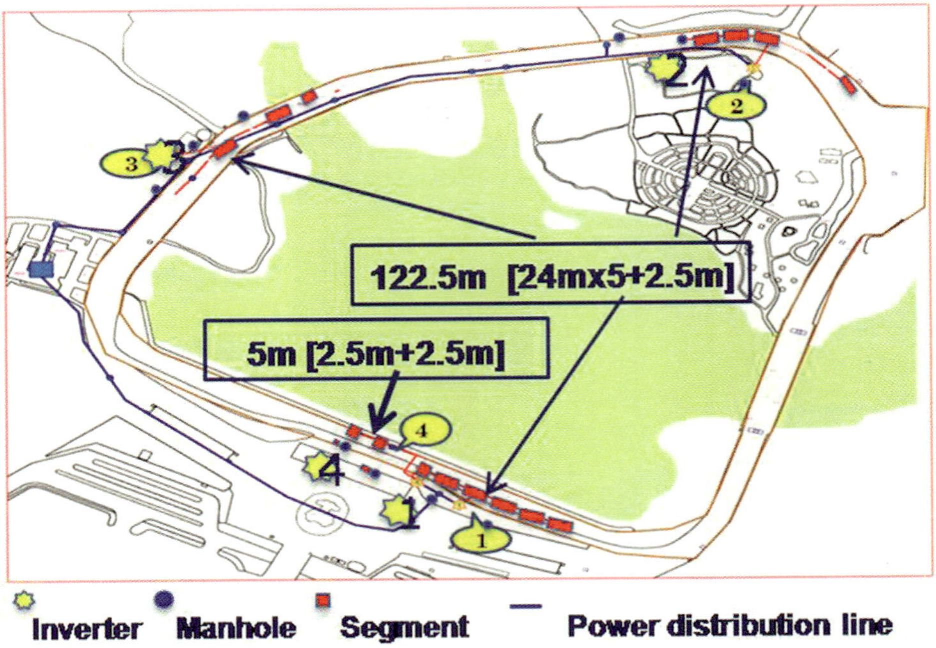


**Figure 4.15** An example of power supply system installed in Seoul Grand Park.

Combining the vehicle with OLEV architecture and the power supply system together, the overall system schematic diagram is shown in Figure 4.16.

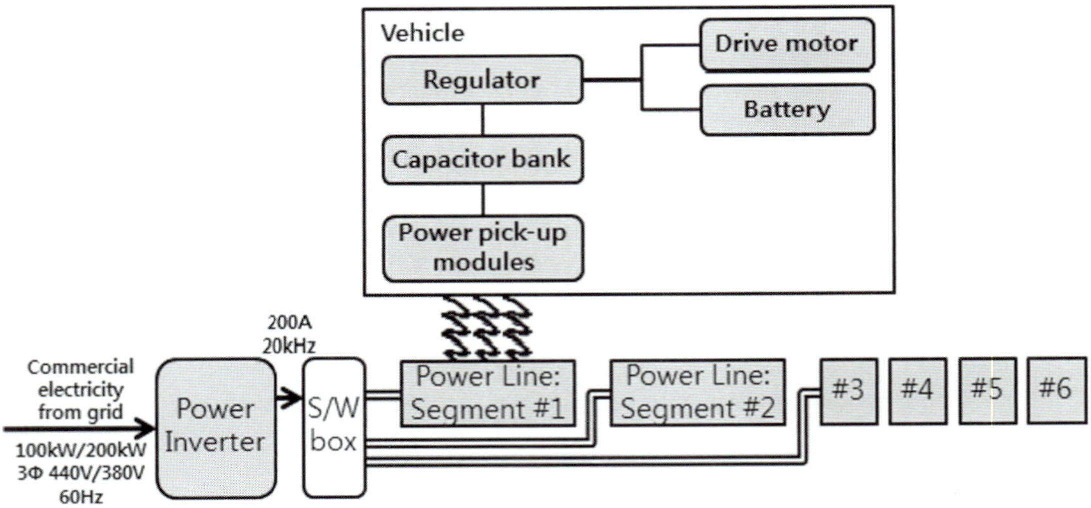


**Figure 4.16** Overall OLEV system architecture: vehicle and power supply system (© 2013 IEEE. Reprinted, with permission, from [4-24]).

### 4.6.2 Electric Vehicle System Architecture

An OLEV can be placed in the category of an electric vehicle because an electric vehicle is defined as a vehicle driven by an electric powertrain with one or more electric motors and a storage medium for electric energy, usually a battery. The unique difference between an OLEV system and a more general electric vehicle is that an OLEV system has a set of pick-up devices installed under the vehicle to collect electromagnetic field energy. A set of electrical devices including rectifiers and regulators is installed to convert and deliver the wirelessly delivered magnetic energy to the applicable electricity of the vehicle for the onboard ESS or directly to drive the motor.

Also necessary within the vehicle is a power control and management system such as a PDU to control the power flow from the electric source of the battery and the supplied power from the road. The control of the electrical power flow and communication of the necessary signals within a vehicle are also incorporated for the proper operation of the vehicle. In Figure 4.17, to show the difference in the system architecture of an OLEV system, a comparison was made between a typical series hybrid electric vehicle in (a) and an OLEV system in (b). Compared with the series hybrid powertrain, the OLEV system does not have an IC engine; thus, there is no tail-out emission.

An example of the physical packaging of the pick-up devices is shown in Figure 4.17 (b). Depending on the electrical load requirement for a vehicle and the power collecting capacity from one unit of the pick-up device, the proper packaging design can be derived by considering the required vehicle ground height. In buses operating inside the KAIST campus, as shown in Figure 4.1 (c) the pick-up power capacity is 75 kW with five sets of pick-up devices with a variable ground height of 15–20 cm, providing the required power to drive full-size city buses with a 120 kW rated and 240 kW peak power ac induction motor.

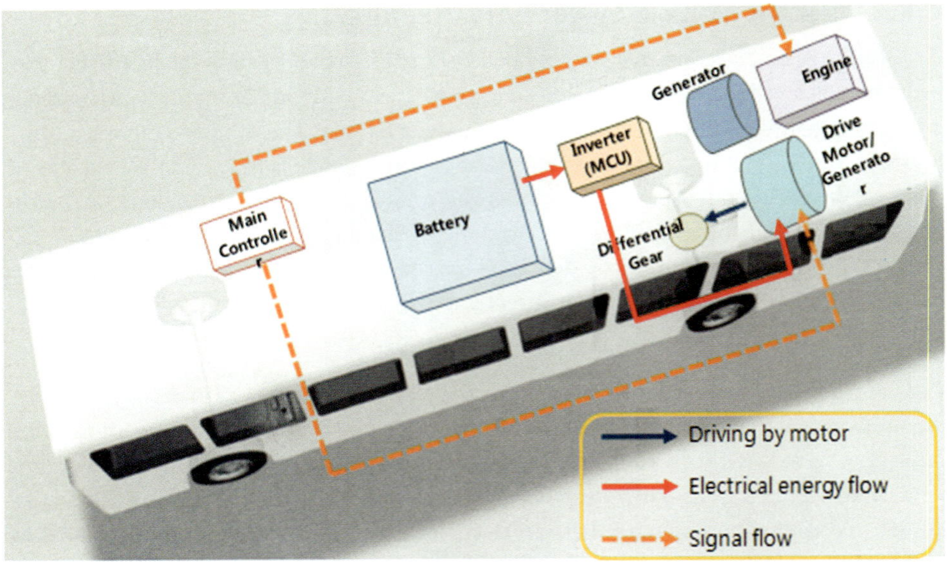


(a)

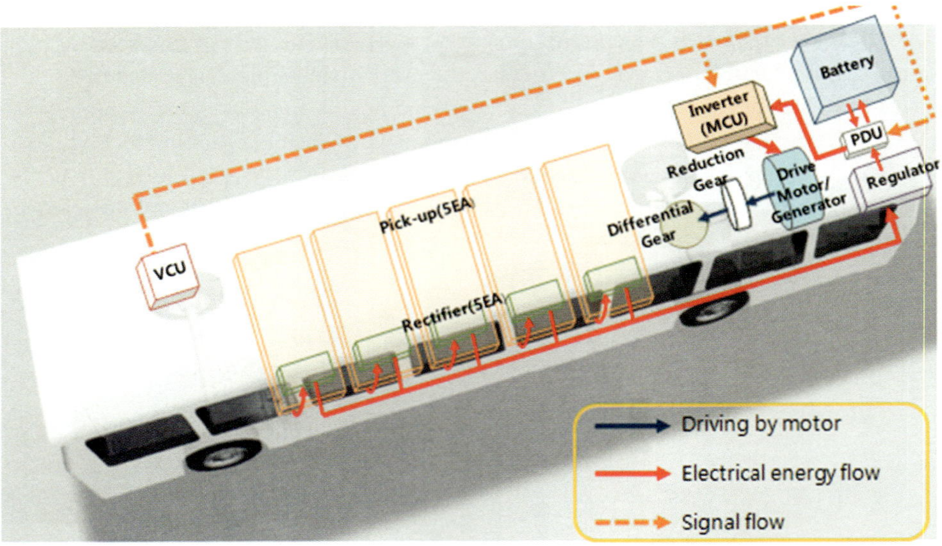


(b)

**Figure 4.17** Schematic comparison of powertrain layouts: (a) Series hybrid system and (b) OLEV system.

## 4.7 Application of SMFIR Technology to Trains

### 4.7.1 Concept

Using a high-capacity, high-efficiency, and low-cost 60 kHz power supply and pick-up system, based on SMFIR technology, the train is wirelessly powered through a power grid built in tracks. A high-capacity pick-up device installed under the train works to collect the magnetic field from on-track power cables and converts it into electric energy for train operation, as shown in Figure 4.18.

**Figure 4.18** Schematic concept of a train system with wireless charging.

### 4.7.2 Main Advantages

To improve their wireless charging capacity and efficiency, a 60 kHz power supply/pick-up system was developed and applied to trains, as shown in Figure 4.19. A wireless power transfer system with an operational frequency of 60 kHz of power supply and pick-up technology was demonstrated at the Osong Train Test Course in June 2013, with its application to catenary-free trains. In the near future, the 60 kHz wireless power transmission technology will be applied to subway trains and high-speed trains. The application of wireless charging technologies to railway systems has the following advantages:

**Figure 4.19** Demonstration of train with 60 kHz wireless charging.

- Strengthening technical competitiveness of the future railroad industry:
  The application of SMFIR technology to railroad systems provides a solution to the problems of conventional wiring, and contributes to an innovative technology development in the future railway industry.

- Reduction of railroad facility maintenance cost and noise:
  As the contactless power supply is adopted, the power is not transferred through wired line, and thus, maintenance costs will be reduced. Also, noise generated due to the catenary during the operation of high-speed trains can be reduced.

- Improvement of urban appearance and reduction in building cost:
  The appearance of streetscapes in urban areas can be improved because there is no need for catenary facilities, such as power poles, and facility building costs decrease because smaller areas are required for railroad sites and tunnels.

## References

4-1 International Panel on Climate Change (IPCC) Fourth Assessment Report (AR4), "Climate Change 2007: Mitigation of Climate Change," IPCC, 2007.

4-2 U.S. Energy Information Administration (EIA), "2002 World Energy Outlook," 2002.

4-3 International Energy Agency (IEA), "Transport, Energy and CO2-Moving toward Sustainability," 2009.

4-4 McKinsey & Company, "Electrifying Cars: How Three Industries Will Evolve," McKinsey Quarterly, Number 3, 2009.

4-5 C. C. Chan, "An overview of electric vehicle technology," Proceedings of the IEEE, vol. 81, no. 9, pp. 1202–1213, Sep. 1993.

4-6 N. P. Suh et al., "Design of On-Line Electric Vehicle (OLEV)," Plenary Presentation, CIRP Design Conference, Apr. 2010.

4-7 I. S. Suh, "Intelligent Wireless EV Fast Charging with SMFIR Technology," Trans. of the Society for Design and Process Science, vol. 15, no. 3, pp. 3–12, Sept. 2011.

4-8 S. E. Shladover, "Roadway-powered Electric Transit Vehicle—Progress and Prospects," Transportation Research Record, no. 1155, pp. 28–36, 1987.

4-9 P. Alto, "Roadway powered electric vehicle project track construction and testing program phase 3D," California PATH Research Paper, UCB-ITS-PRR-94-07, 1994.

4-10 KAIST OLEV Project Center, "OLEV Project Overview," KAIST Internal Document, Apr. 2010.

4-11 I. S. Suh, "KAIST On-Line Electric Vehicle as a Future Transportation Technology with Low Carbon and Green Growth," International Forum on Public Transportation Policy hosted by KOTI, Korea, Mar. 2010.

4-12 Jaegue Shin, J. G. Shin, S. Y. Shin, Y. G. Kim, S. Y. Ahn et el., "Design and Implementation of Shaped Magnetic-Resonance-Based Wireless Power Transfer System for Roadway-Powered Moving Electric Vehicles," IEEE Trans. On Industrial Electronics, vol. 61, no. 3, pp. 1179–1192, Mar. 2014.

4-13 I. Suh et al., "An Introduction of On-Line Electric Vehicle System and its Pilot Application in the Seoul Grand Park," 17th ITS World Congress, Busan, 2010.

4-14 I. Suh et al., "Application of SMFIR Technology to Urban Transportation," CIRP Design Conference, 2011, Daejeon, Korea.

4-15 ICNIRP Guidelines, "Guidelines for limiting exposure to time-varying electric, magnetic, and electromagnetic fields (UP TO 300 GHz)," Health Physics, vol. 74, no. 4, Apr. 1998.

4-16 A. Kurs, A. Karalis, R. Moffatt, J. D. Joannopoulos, P. Fisher, and M. Solja i , "Wireless Power Transfer via Strongly Coupled Magnetic Resonances," Science, vol. 317, no. 5834, pp. 83–86, Jul. 2007.

4-17 C. S. Wang, O. H. Stielau, and G. A. Covic, "Design considerations for a contactless electric vehicle battery charger," IEEE Trans. Ind. Electron., vol. 52, no. 10, pp. 1308–1314, Oct. 2005.

4-18 G. A. Covic, J. T. Boys, and M. L. G. Kissin, "A Three-Phase Inductive Power Transfer System for Roadway-Powered Vehicles," IEEE Transactions on Industrial Electronics, vol. 54, no. 6, pp. 3370–3378, Dec. 2007.

4-19 T. Imura and Y. Hori, "Maximizing air gap and efficiency of magnetic resonant coupling for wireless power transfer using equivalent circuit and Neumann formula," IEEE Trans. Ind. Electron., vol. 58, no. 10, pp. 4746–4752, Oct. 2011.

4-20 D. Ahn and S. Hong, "A study on magnetic field repeater in wireless power transfer," IEEE Trans. Ind. Electron., vol. 60, no. 1, pp. 360–371, Jan. 2013.

4-21 T. C. Beh, M. T. Kato, S. Oh, and Y. Hori, "Automated impedance matching system for robust wireless power transfer via magnetic resonance coupling," IEEE Trans. Ind. Electron., vol. 60, no. 9, pp. 3689–3698, Sep. 2013.

4-22 L. Chen, S. Liu, Y. C. Zhou, and T. J. Cui, "An optimizable circuit structure for high-efficiency wireless power transfer," IEEE Trans. Ind. Electron., vol. 60, no. 1, pp. 339–349, Jan. 2013.

4-23 S. Chopra and P. Bauer, "Driving range extension of EV with on-road contactless power transfer-A case study," IEEE Trans. Ind. Electron., vol. 60, no. 1, pp. 329–338, Jan. 2013.

4-24 J. Kim, J. Kim, S. Kong, H. Kim, I. S. Suh, N. P. Suh, D. H. Cho, J. Kim, and S. Ahn, "Coil design and shielding methods for a magnetic resonant wireless power transfer system," in Proc. IEEE, vol. 101, no. 6, pp. 1332–1342, Jun. 2013.

4-25 I. S. Suh, "On-road Electrification for Optimized Power Supply in OLEV Application," Trans. of the Society for Design and Process Science, vol. 15, no. 3, pp. 13–27, Sep. 2011.

4-26 P. Kemper and I. S. Suh, "Implementation cost comparison of different types of EV energy replenishment technologies for public transit bus systems," Transportation Research Record, 2014.

# Chapter 5
# Wireless Power Transfer Technical Issues and Challenges

This chapter will discuss the key technical issues concerning the deployment and operation of wireless charging of PEVs. The primary challenge is communications. Wireless charging, by its very nature, is reliant on private and secure communications that reliably communicate vehicle battery pack status and charge rate demand to the grid-side converter so that the battery is charged according to a specified program of charge rate as a function of the endogenous variables and exogenous variables. Endogenous variables include SOC, voltage, and temperature; exogenous variables include charging pad alignment, gap variations, and manufacturing tolerance of the power transfer components. These issues and the challenges they pose are the topic of this chapter.

## 5.1 Vehicle to Infrastructure Communications

Wireless charging means exactly that—completely wireless power transfer and communications. The main benefit of wireless charging is convenience, as there is no contact potential of the vehicle relative to the ground, no ground currents, dedicated and secure communications, and full autonomy. These are very exciting benefits to the customer because now the electric car can essentially operate itself. Through V2I communications, the charging process can be autonomous and thereby put the vehicle owner's mind at ease that charging will be performed without need for cables and plugs and that he or she will not even need to remember when charging must be done.

Figure 5.1 depicts the major elements of wireless power transfer charging of an electric vehicle high-voltage battery pack and how V2I communications form the outer loop in power flow regulation. In this architecture the power control is done grid-side, much like DC fast charging, so vehicle complexity is minimized and only high-frequency ac current rectification, filtering, and possibly some power conversion is performed at the WPT secondary side.

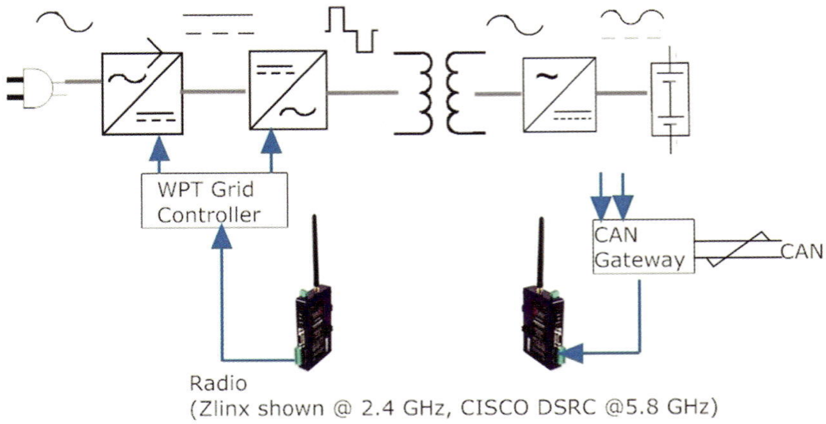


**Figure 5.1** V2I communications in the WPT outer regulation loop.

Figure 5.1 shows four major functional blocks between the utility service and the vehicle HV battery. These functions are explained here and constitute the minimal set of elements that comprise WPT charging.

1. A grid input AFE or PFC stage that performs double duty as grid-side power quality assurance and as a variable dc link voltage source. The power factor is maintained above 98% at 25% power and higher. Also, line total harmonic distortion (THD) is maintained below 5%. The output of this stage, as shown graphically, is variable dc voltage.
2. The high-frequency power converter, typically a power electronic H-bridge or back-to-back mid-point converter, acts as a voltage source and synthesizes a quasi-square wave voltage that is applied via a high frequency isolation transformer to the series-tuned primary of the coupling coil. As shown graphically, the quasi-square wave has an adjustable duty ratio that can be used to fine tune the overall voltage control of the power flow. The isolation transformer, not shown but part of the HF inverter, also has dual roles: a) providing galvanic isolation between the utility ac line and the HF cable connection to the primary WPT coil and b) providing matching between the HF bridge operating at high voltage and the relatively low voltage and high current of the primary coil. This latter feature brings an added benefit in realizing the highest efficiency in the HF inverter by minimizing its current for a given power rating.
3. The coupling coil pair, shown graphically as a transformer symbol, is tuned by a series capacitor to the desired operating center frequency. This tuning effectively cancels out the loosely coupled transformer primary leakage inductance, thereby reducing its impedance to the application of high primary current. It is primary

current that establishes the coupling field (i.e., magnetic flux) necessary to induce voltage into the more remote secondary coil on the underside of a vehicle under charge. Because of the high-frequency nature of the excitation current, the coupling coils are wound with Litz cable. This is necessary to minimize the ratio of ac resistance to its dc resistance and maximize the coupling efficiency.

4. Stage 4 is the secondary side tuning capacitor, rectifier, and filtering network. The secondary tuning capacitor acts much like the primary tuning capacitor; its main function is to offset the voltage drop introduced by the secondary leakage inductance and thereby maximize the voltage presented to the rectifier stage and from there on to the vehicle HV battery. Filtering is needed again to smooth the HF full-wave rectified current before it reaches the battery. Some ripples will be present in the secondary rectified dc, as the graphic shows. The battery symbol is the PEV HV traction battery.
5. Communication is provided by radio modem, such as the Zlinx units shown in Figure 5.1, or by more up-rated radios, such as CISCO dedicated short range communications (DSRC). On the vehicle side, certain signals on the WPT secondary and from the HV battery must be communicated back to the grid converter for the purpose of power flow regulation. As a minimal set, the secondary rectified voltage, the battery voltage, battery SOC, and battery temperature and other temperatures must be monitored. For example, prior to closing the WPT contactor that connects the rectified output to the HV battery, the filter capacitors must be pre-charged by the primary. To do so, the communications requests some low level of primary output power, then monitors the WPT dc voltage and battery voltage, and, when the difference is within an acceptable level, the contactor is closed. This avoids any impulse of current that could damage the WPT components.

Additional information on the HV battery is monitored and communicated to the grid-side converter. Consider how charge rate is controlled. Local sensors monitor WPT filtered output current, voltage, and temperature. These local measurements are shown being input directly into a CAN gateway module for transmission to the grid controller. Multiplication of the charge current with battery voltage is the charging power level that must be commensurate with battery SOC and its temperature, SOC(T). Battery status such as SOC and temperature are available from the battery management unit over the CAN network. This information is essential to control during charging and, because this is a stationary setting, the radio modem message latency presents no real issues or challenges. Although, to be fair, the message repetition rates will be assigned according to priority, such as charge rate and voltages being more urgent than battery pack temperature. Beyond this are the administrative controls needed to ensure safe and robust charging in an autonomous setting. For example, the grid side unit will not enter into charging mode until messages are sent stating that the vehicle is in a key-OFF state, in park, and inhibited from drive-away. The CAN gateway, or other WPT controller modules, would also ensure that the conductive charger function is disabled and some manner of display notification is made that conveys this information. In addition, object detection of inanimate or biological intrusion into the WPT active zone is required, and, should metal foil or a small animal be detected on or near the primary coil, the charging would be inhibited until the object was cleared. During this handshaking interval, the V2I link to utility services may also report to the charger pertinent grid information such as TOU and burden rate limitations due to grid capacity and loading.

The radio modem on the WPT primary side connects directly into the wireless charger controller, for example, via direct RS232, RS485, or similar protocol. A message calling for an increase in charging power, for example, would direct the AFE to increase the voltage set point on the PFC voltage loop and thereby increase the amplitude of the primary quasi-square wave voltage according to Equation 5.1. Note that adjustment of HF inverter rail voltage $U_{d0}$ has the most direct effect on the source voltage, whereas duty ratio **d** is less pronounced until it reaches relatively low values. The inverter operating frequency is set by the grid controller and will be standardized to a center frequency in an available band within the 10 kHz to 100 kHz available frequency allocation. It is found that as duty **d** is reduced from 0.8 to below 0.5, WPT input reactive power increases quickly. Therefore, duty should be maintained in the range 0.7 < **d** < 0.9 for the lowest reactive burden on the HF inverter.

$$U_s(t) = \frac{4U_{d0}}{\pi}\sin\left(d\frac{\pi}{2}\right)\cos(\omega t) \qquad V_{rms} \tag{5.1}$$

The amount of power that can be transferred between the primary and secondary and to the vehicle battery in a wireless charger is strongly linked to the open circuit voltage at the secondary for a specified primary current $I_p$, and the physical gap between the coils, $z$, as discussed by Covic [5-1] and stated as Equation 5.2 for reference. In this expression, the open circuit voltage results from the mutual inductance between the primary and secondary coils—that is, as a function of coil spacing $z$, according to the coil pairs coefficient of coupling $k(z)$. The factor $Q_S$ in Equation 5.2 is the quality factor of the secondary coil due to its resistive loss components and will have a nominal value in the range of 3 to 5. This means that the highest possible short-circuit capability is desired because $I_{sc} = k(z)I_p$.

$$P_{max} = U_{oc}I_{sc}Q_s \qquad W \tag{5.2}$$

Figure 5.2 illustrates one possible AFE according to [5-2] and [5-3], wherein the rectified current from an ac grid connection is actively rectified via the semiconductor switch and auto-commutating diode arrangements Tb1–D1 through Tb4–D4. HF inverter link voltage $U_{d0}$ can be regulated below the rectified line voltage by the cascaded buck converter stage, C1, Tb5, D5, and L3. The HF inverter is shown shaded and interfaced to the primary tuned circuit via the isolation transformer. Note that the primary coil pad enclosure should be earth grounded to the grid converter case ground to avoid any contact potentials.

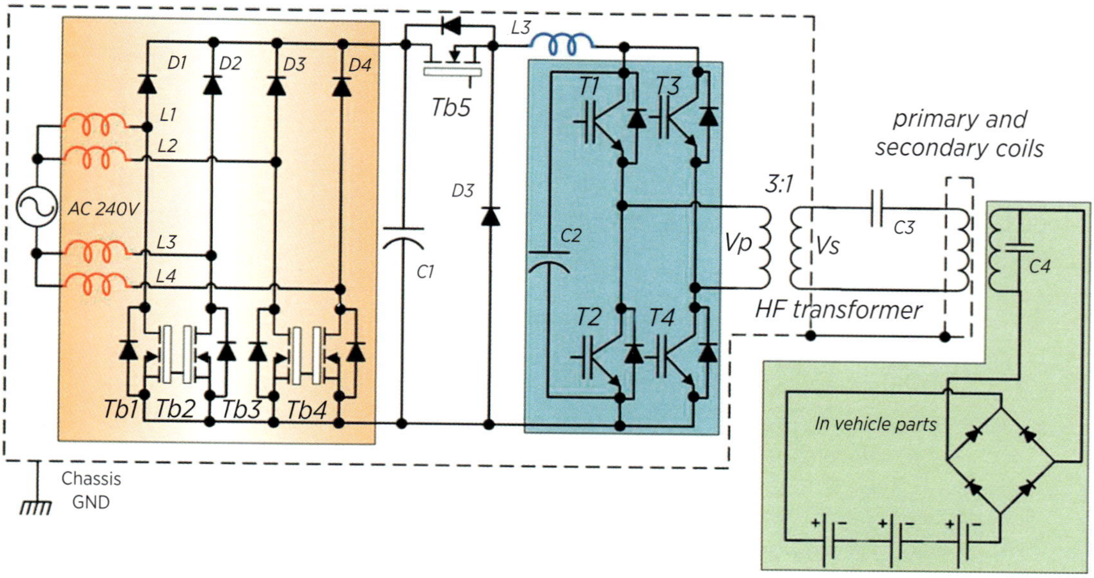


**Figure 5.2** WPT active front end, HF inverter, and coupling coils to vehicle-side rectifier (graphic courtesy of Dr. Lixin Tang, ORNL).

The primary side radio modem shown in Figure 5.1 as interfacing the WPT grid controller, in effect, controls AFE power transistors Tb1–Tb5 via matching gate drivers to adjust the HF inverter rail voltage. The HF inverter is similarly controlled via matching gate drivers to power switches T1–T4. Therefore, rail voltage $U_{d0}$ given in Equation 5.1 is controlled by this first set of power transistors, and both duty **d** and angular frequency $\omega$ are set by the second group of power transistors. An additional current regulation loop on the AFE is used to maintain grid input power factor, PF > 0.95. With proper switching frequency of the AFE transistors and passive components, the input total harmonic distortion can be maintained below 5%.

## 5.2 Alignment to Primary Charging Pad

Alignment is a challenging aspect of wireless charging, as the primary and secondary coils must be brought into relatively good proximity for efficient and effective power transfer. Refer to Figure 5.3 for an illustration of a central mounting location of the secondary coil on the vehicle. Notice that, in a center-mounted location, the secondary coil will reside immediately below the driver and passenger seats. A metric of concern in wireless charging is the distance from the center of the primary coil to the side of the vehicle where someone may be standing and potentially be exposed to high-frequency magnetic fields, albeit at very low strength. This distance is labeled $y_0$ for future reference in discussions about emissions. Note also that nominal ground clearance of light-duty vehicles yields an effective coil gap of $z_0$ ~160 mm.

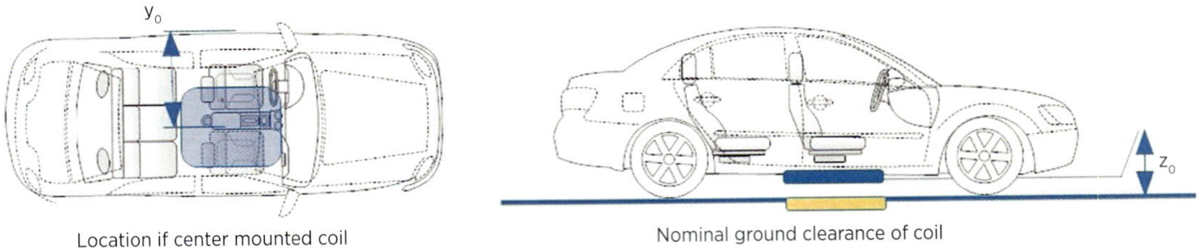


**Figure 5.3** WPT central mounting location beneath the vehicle.

Issues arising from misalignment include its influence on coupling efficiency and fringe fields, as well as emissions in both magnetic and electric fields. Discussion on electric and magnetic field emissions will be taken up in a later subsection. The challenges posed by secondary coil mounting and alignment are several:

- Impact of secondary coil mounting on vehicle ground clearance and avoidance of damage due to contacting obstacles such as speed bumps and road debris
- Environmental sealing of the secondary coil and cable protection and routing—for example, sealing to prevent water intrusion (or submersion) and contamination by dirt, dust, and lubricants
- Location of hard points beneath the vehicle chassis to affix the coil to, and what impact this will have on the vehicle floor pan integrity, especially if nonmetallic
- Measures that need to be taken to further shield the passenger compartment from magnetic fields to address future wireless charging in a quasi-stationary case such as charging at stop lights, on roadway lay-bys, and other occasions when the driver and/or passengers are present

Figure 5.4 expands on the coil mounting location shown in Figure 5.3 as misalignment in the vehicle longitudinal dimension $\Delta x$, and in the vehicle lateral dimension $\Delta y$. Wireless chargers having proper flux guide designs can easily accommodate misalignment in $(x,y)$ of up to 20% individually, but, in combination, the challenge is maintaining the power transfer magnitude when the overlap area drops below 80% of nominal or fully aligned condition. For example, in square geometry coils, a compound misalignment of just 11% in $\Delta x$ and $\Delta y$ reduces the overlap area to 80%.

For these reasons, various alignment techniques have been explored to ensure the vehicle is parked properly over the charging pad. Options include visual alignment using a chassis-mounted camera, much like a back-up camera used in some vehicles, or radio frequency ID (RFID) and triangulation techniques, or using the pads themselves and seeking alignment using its magnetic field. Simple alignment methods could consist of tire curbs and chock blocks to first align laterally and then stop at longitudinal alignment. However, that option only works for a specific class of vehicle, as smaller or larger chassis vehicles will not fit the mechanical stops. Some advocates of wireless charging recommend mechanical systems having articulated members to locate the charging pad, but that introduces considerable cost.

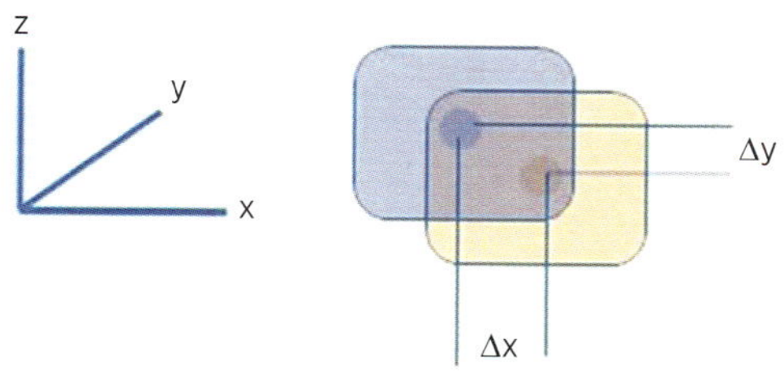


**Figure 5.4** Definition of misalignment in vehicle longitudinal (*x*) and lateral (*y*) dimensions.

## 5.3 Challenge of Gap Variations

The physical separation between a primary charging pad and vehicle-mounted secondary pad is the air gap clearance $z_0$ between the two enclosures, whereas the magnetic gap is the distance, $z = 0$ between the respective conductor coils. For most light-duty vehicles, the nominal clearance $z_0$ is on the order of 160 mm; therefore, wireless charging must be validated for clearances in the range of $d = 125$ mm to $d = 200$ mm or more. The range in magnetic gap must cover the cases of vehicle loading and primary coil location whether on the surface, embedded flush with a surface, or beneath the surface. The first two primary locations would be representative of wireless chargers in a residential garage, parking structure, or parking lot. The latter location is more representative of placement on a highway. Figure 5.5 illustrates the nominal ground clearance of a WPT installation in a parking garage.

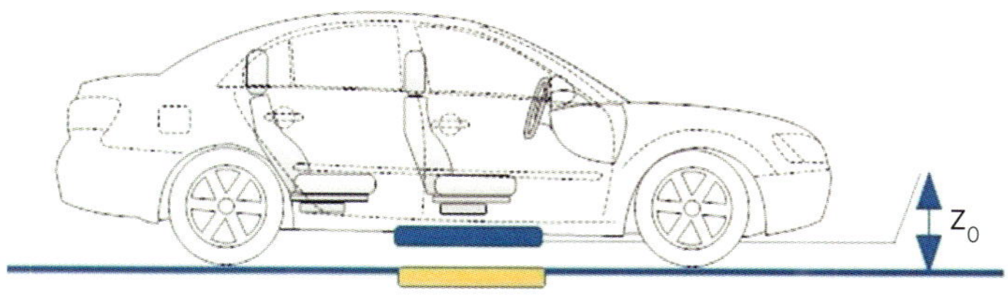


**Figure 5.5** Primary to secondary coil clearance.

From the perspective of the magnetic field, the space between the coils must be magnetized to couple power from the primary to secondary. The larger the magnetic gap, the more magnetizing current is required to magnetize it to a level sufficient for effective power transfer. Figure 5.6 is representative of the magnetizing field in which a considerable portion of the field simply does not link the secondary coil and is a leakage field. Magnetic leakage exists within the coil space where flux lines loop the primary conductors and not the secondary conductors, but do not reach beyond the confines of the coil structure, whereas fringe fields are those magnetic flux lines that do reach outside the coil confines and contribute to emissions that must be minimized.

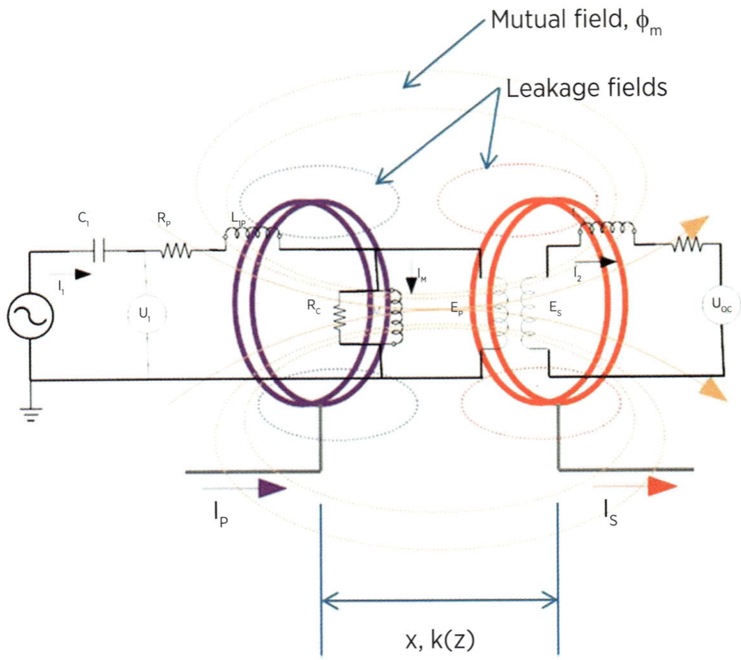


**Figure 5.6** Description of WPT magnetic fields.

The following key observations can be made from Figure 5.6:

- The mutual field magnetizes the gap and facilitates power transfer from grid to vehicle
- Primary current $I_p$ establishes the total field, mutual + leakage magnetic flux lines
- Leakage fields are highly gap ($z$) dependent and diminish the coil coupling coefficient $k(z)$
- The mutual field at secondary coil produces a voltage $E_m$ that supplies charging current
- Primary tuning capacitor $C_1$ resonates out primary leakage, thereby facilitating high primary current $I_p$ from source $U_S$ and, from it, power transfer

The equivalent circuit overlaid on the WPT coils and field lines in Figure 5.6 depicts the electrical analogy of the electromagnetic processes. As shown left to right, the source $U_S$

energizes the primary coil with current $I_p$. The primary-series-connected tuning capacitor resonates out the primary leakage inductance, thereby facilitating higher primary current for a given level of voltage $U_S$. The coupling coils are modeled as the "T" equivalent of a transformer having magnetizing branch inductance (i.e., inductance of the mutual field M), primary and secondary leakage inductances, an ideal transformer to model the case of non-equal turns between primary and secondary, and a core loss term $R_c$ to account for energy loss in the magnetic components used to focus and shield the magnetic fields. This loss term also includes additional losses due, for example, to covering material on the primary pad such as concrete or asphalt in cases where it is used (e.g., on highways).

The coupling coefficient $k(z)$ is shown having $z$-axis variation according to the vehicle ground clearance, as was noted. In the context of the electromagnetic variables, this coupling coefficient is defined as Equation 5.3, the ratio of that portion of primary magnetic flux that links the secondary $\phi_m$ to the total amount of flux emanating from the primary coil $\phi_p$ due to its turns and conductor current.

$$k(z) = \frac{\phi_m}{\phi_p} \tag{5.3}$$

For the reader familiar with transformer theory, this coupling coefficient can be written in terms of the inductances shown in Figure 5.6 as Equation 5.4. In the equation, the magnetizing inductance is $M$.

$$k(z) = \frac{M(z)}{\sqrt{L_p L_s}} \tag{5.4}$$

Referring again to Figure 5.6, the reader will note that maximizing power transfer requires a secondary winding potential $E_p$ that is sufficiently high to match the load voltage (i.e., HV battery voltage). Figure 5.6 is generic in that non-unity turns ratio of the WPT coupler is accommodated (e.g., $E_p \neq E_s$), but for unity turns ratio these can be replaced with a potential across the magnetizing field of the coupler $E_m$. This relationship is given as Equation 5.5, and shows how important maximizing the coupling coefficient is, including the role of frequency $\omega$ and primary current $I_p$.

$$E_m(z) = j\omega M(z) I_p \tag{5.5}$$

Figure 5.7 illustrates an experimental coupling coil set having the parameters listed in Table 5.1 and with coupling coefficient $k(z)$ experimentally measured by three different techniques. Note that the coupling coefficient varies reciprocally with gap $z$, and that it falls in the range of $0.15 < k(z) < 0.3$ when $125 < z < 200$ mm.

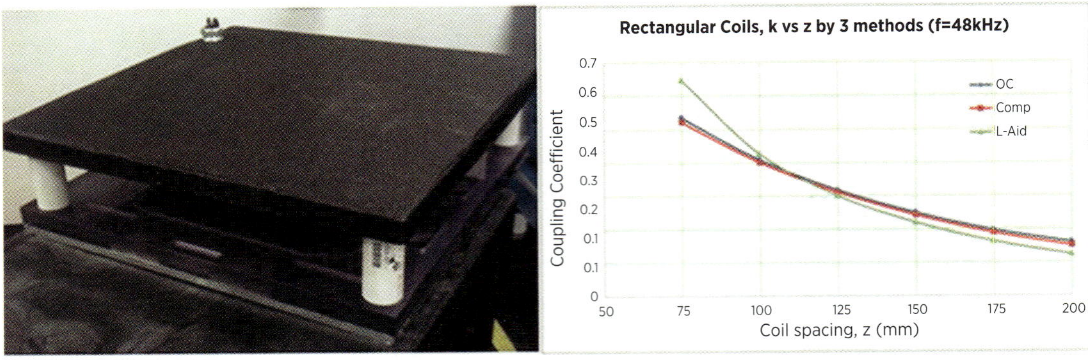


**Figure 5.7** Experimental coupling coil pair and measured coupling coefficient $k(z)$ (coupling coil courtesy of ORNL).

**Table 5.1 Coil Parameters for Experimental Coil Shown in Figure 5.7**

| | | | | | | | |
|---|---|---|---|---|---|---|---|
| *Coil width, max* | **a** | *533* | *(mm)* | *Coil length, max* | **b** | *558* | *(mm)* |
| *Equivalent radius* | $r_{eq}$ | *300* | *(mm)* | *Coil area* | $S$ | *0.297* | $m^2$ |
| *Nominal working gap,* *D/4 =* $r_{eq}/2$ | $d$ | *150* | *(mm)* | *Coil turns (primary & secondary)* | $N_c$ | *11* | *(#)* |
| *Coil DC resistance* | $R_{dc}$ | *22* | *(mΩ)* | *Coil inductance* | $L_p$ | *122* | *(μH)* |
| *Coil AC resistance, f* = 25 kHz | $R_{ac}$ | *38* | *(mΩ)* | *Coil AC resistance, f* = 48 kHz | $R_{ac}$ | *64* | *(mΩ)* |

In Figure 5.7, the agreement in coupling coil measurement using three different techniques is very good. The most direct method, that of measuring open circuit (OC) voltage for a prescribed magnitude of primary current, gives results nearly identical to those of the method of compensating this measurement with short-circuit current. The method of measuring the coupling coefficient is shown as L-aid and requires a laboratory instrument to measure inductance when the primary and secondary coils are magnetically coupled.

As the reader would expect, the coupling coefficient increases quickly as the gap is reduced to zero and, in fact, becomes approximately unity at zero gap. Take as an example a specified frequency and voltage $E_m$ in Equation 5.5 and the range variations for gap and coupling coefficient noted in Figure 5.7. When the gap is 125 mm, $k(z = 125) = 0.3$ and current $I_p$ = 1pu (per unit) is applied to the primary, then when the gap increases to $z$ = 200 mm, the primary current $I_p$ = 2pu since coupling coefficient was reduced to half its previous value. This is the reason the transmitting power across large gaps requires larger-diameter coils. As noted in Table 5.1, a reasonable working gap is 1/4th the coil diameter $D$.

## 5.4 Control of Charging Process

Wireless charging can be controlled from the primary side only (as in dc fast charging), at the secondary (i.e., vehicle) side only (as in level 1 and 2 conductive charging), or as a combination of the two. The challenge going forward in this exciting application area is standardization, and charging regulation type is no small issue. Therefore, in this section, only primary (i.e., grid-side) regulation is considered. The authors in [5-4] describe grid-side WPT charging according to Figure 5.8, where vehicle HV battery variables are monitored, charging power delivered is calculated, and the communications channel is in the power regulation outer loop. The function of this communications loop in the regulation process was discussed previously in section 5.1.

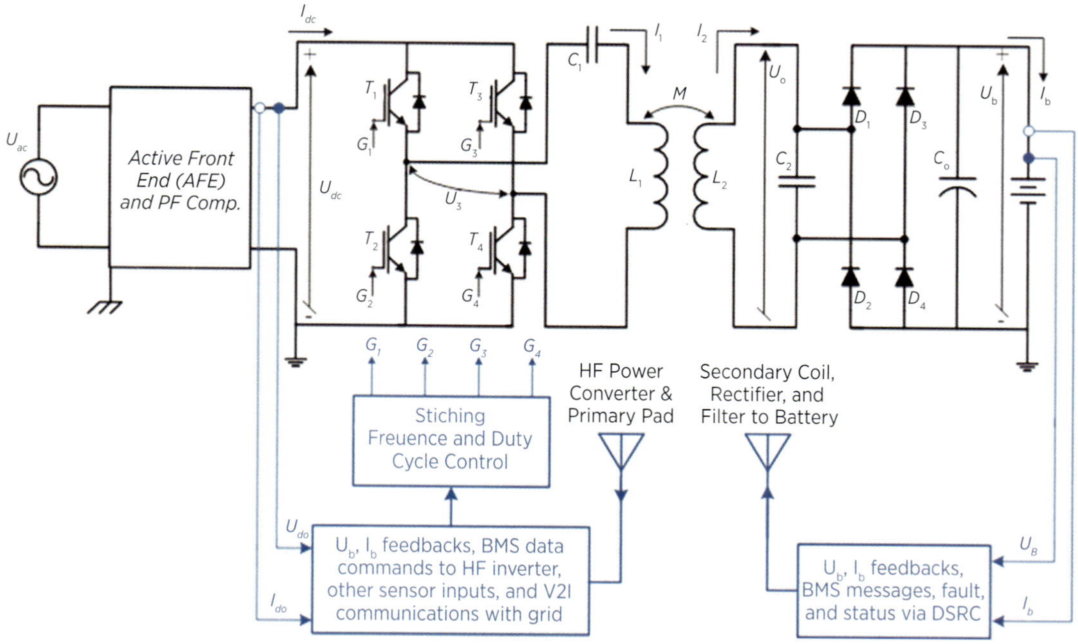


**Figure 5.8** Representative primary side only control of wireless charging.

The function of the power flow regulator is to maintain equivalent constant current (CC) and constant voltage (CV) charging as a function of the battery pack SOC. These charge functions are implemented, as shown in Figure 5.8, based on monitored WPT charge parameters at the battery, $U_b$ and $I_b$, and their product $P_{bat} = U_b \cdot I_b$ for real power delivered to the battery terminals. The outer regulation loop, of which the communications channel is a part, sends this charging power message to the grid converter controller where it is compared to the AFE output power, $P_{AFE} = U_{d0} \cdot I_{d0}$. The power electronics in the WPT AFE ensure proper power quality at the grid connection and also the proper level of real power to the HF H-bridge. It is not necessary to account for losses in the power transmission cascade because the power delivered is under closed-loop control. However, it is necessary to properly regulate the power flow control variables $U_{d0}$, $\omega$, and **d**, in that order.

Figure 5.9 illustrates the charging sequence followed for a lithium-ion pack. Charging takes place over the following three distinct periods to bring the pack up to its top of charge: 1) pre-charge, 2) constant current charge, and 3) constant voltage taper charge. Depending on its SOC and temperature, the pre-charge interval may need to be prolonged to bring the battery cell pack into a SOC range where higher-rate charging current can be tolerated. During CC charging, the HV battery SOC can be increased at a rate determined by its charge current limit obtained from the vehicle battery management unit via CAN and the grid-side converter power delivery if limited by utility TOU and any charging restrictions due to capacity limitations. The CC interval is also labeled "fast-charge" if the vehicle manufacturer offers this high-rate charging feature.

During the final charge period, based on the battery status obtained from the vehicle battery management unit, the charger will enter into taper charge, or CV mode at a charge rate that is restricted for the final 20 to 30% SOC, when the pack SOC transition passed 70 to 80%. The CV period is very important because overcharging will result in battery heating and potential damage.

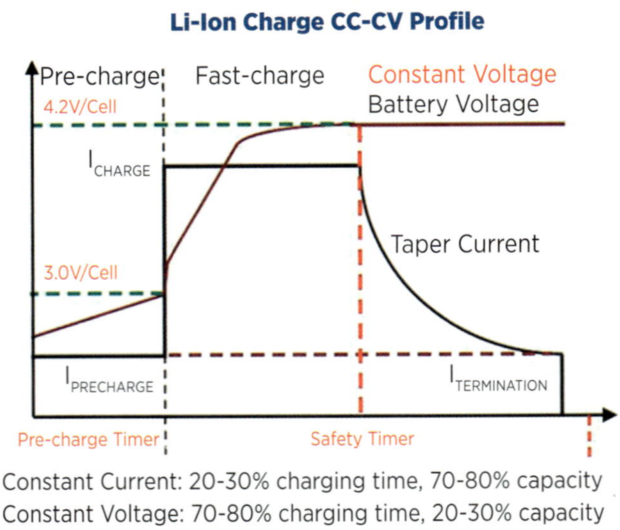


**Figure 5.9** Representative charging profile for lithium-ion battery (graphic courtesy of Dr. Burak Ozpineci, ORNL).

## 5.5 Obstacle Detection

For all its convenience and autonomy, a wireless charger does have issues with conductive objects in the active zone. Because techniques for obstacle detection can be proprietary and manufacturer-specific, it will suffice to illustrate the concern and to list technologies that may be employed to detect objects, whether biological, conductive, or insulating. A short list includes the following:

- Monitoring abnormal input power demand not matching power delivered to the HV battery
- Periodic impedance tests of the primary pad
- Use of motion detectors and optical means such as camera and image processing

Figure 5.10 is presented here to illustrate one potential obstacle that may be encountered, and that is an extension cord on the primary charging pad. If the cord ends are unplugged, then voltage will be induced in the conductors of the cord, but no current will flow because there would be no closed path for current to follow. However, the cord prong end and receptacle would have some magnitude of HF ac voltage induced and present.

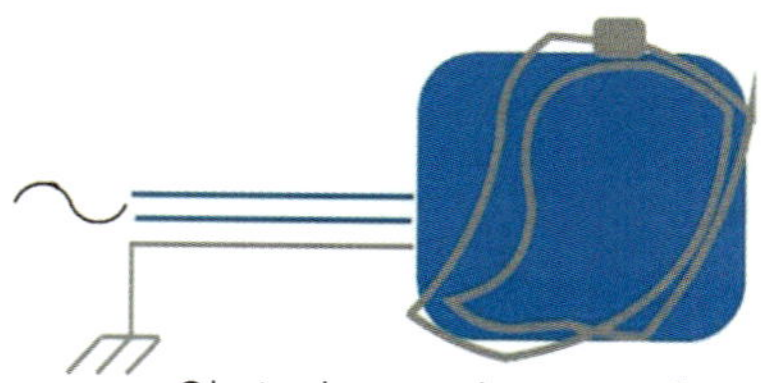


**Figure 5.10** Illustration of an extension cord on a primary charging pad.

The case that presents an issue is when the cord ends are plugged together so that the extension cord becomes a shorted secondary of a transformer. In this case, a short circuit will be reflected into the WPT primary circuit, and excessive currents will flow that may overload the HF inverter. A current sensor for primary current will detect the abnormal situation and signal a rapid shut-down of charging operations.

## 5.6 Emergency Shut-Down

During wireless charging, all the power transmitted from the grid to the vehicle ESS flows across an open gap in the WPT active zone. This means that the intrusion of foreign objects, such as conductive debris and biological entities, is possible; this is an issue that must be addressed. Object detection, discussed earlier, is the first layer of defense, the second being what action will be taken and how fast it can be accomplished. This topic was taken up in [5-5] and explained using the WPT installation diagram shown in Figure 5.11 for a PHEV illustrating the utility connection and energy metering through the wireless charger into the vehicle and its energy storage system. Charge indication and alignment are some of the data displayed to the vehicle operator.

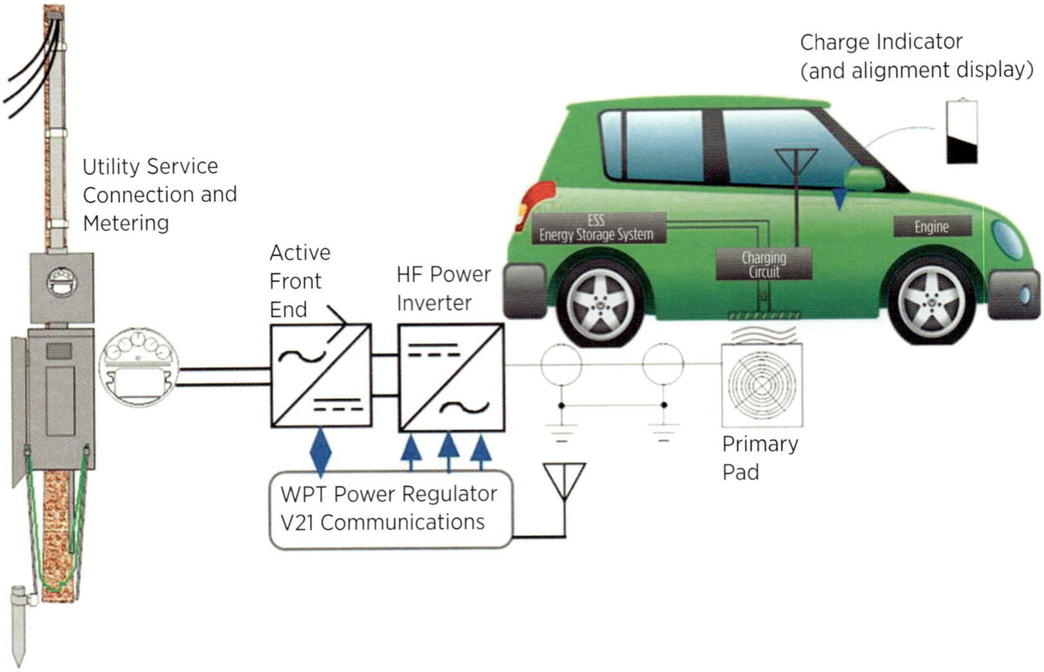


**Figure 5.11** WPT installation with V2I communications.

For electrical safety, the grid connection provides an earth ground at the service entrance. The WPT primary pad enclosure is shown grounded to the WPT base unit, and this unit enclosure (not shown) is, in turn, bonded to the earth ground. Therefore, no contact potentials exist on the charging equipment, and no ground potential exists due to a single earth connection. All utility connections must be made by a qualified electrical contractor and the installation inspected and approved by the authority having jurisdiction (AHJ).

Before an AHJ will approve the installation, the WPT charger must be certified as meeting applicable electrical codes and adhering to SAE J2954 standard and to UL 2750 test validation. WPT charger commissioning will, therefore, require the use of a standard receiver (secondary) coil and use of a field installation test and measurement validation procedure.

Given a properly installed and commissioned wireless charger, the issue at hand is how would this system respond to either a system malfunction or some object intrusion into the active zone? Referring to Figure 5.11, the question then is how fast can a fault (or intrusion) in the vehicle ESS or related system be communicated to the WPT base unit to trip it off line? To address this challenge, the follow sequence of events is considered and posed as a representative response to a fault condition. At time $t = t_0$, the BMS issues a stop charge message over CAN. After a short communications delay of $\varepsilon$ over CAN from BMS to the gateway module, a local command is issued to disengage the WPT relay.

The command to disengage the WPT relay shown as step 2 in Figure 5.12 is executed over the CAN gateway module input/output (I/O). During this step, the same message to inhibit power transfer is communicated to the V2I radio modem from the CAN gateway module. There will be some short delay $\varepsilon$ over this interface (i.e., RS232 or RS485) plus somewhat longer V2I communications delay $\delta$ before the WPT base unit is disabled, as described in detail in Figure 5.13, where the message routing is picked up at the V2I link. During this communications channel message time of perhaps 10 ms, which is shown as step 4, the vehicle-side WPT relay will be turned off and charging power to the battery will be interrupted, the same process in step 3.

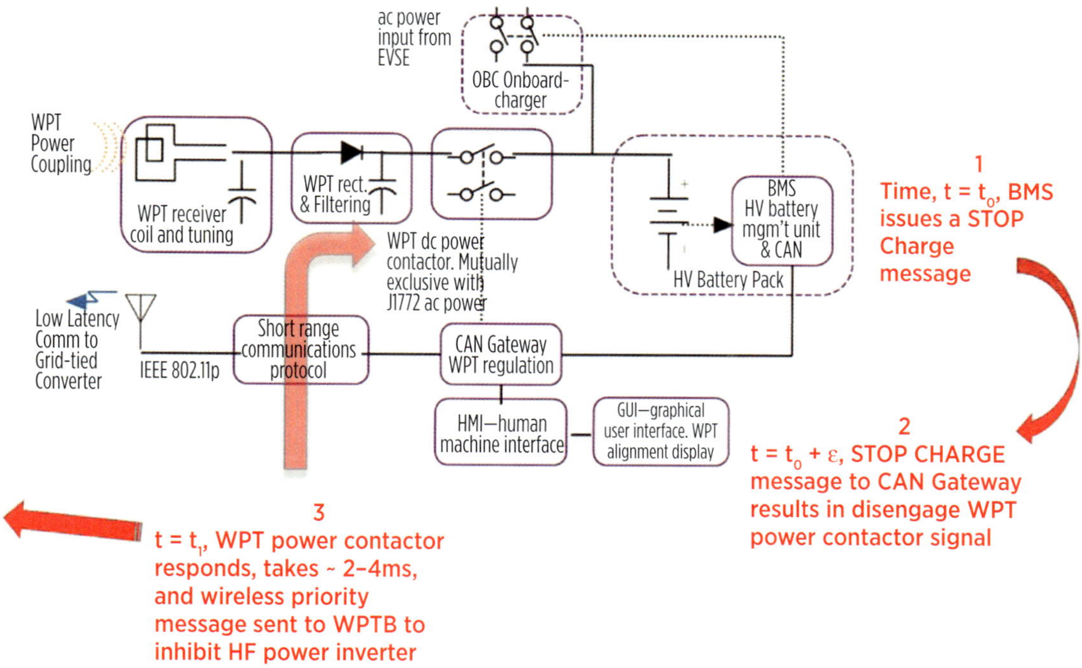


**Figure 5.12** WPT fault detection and first-level response.

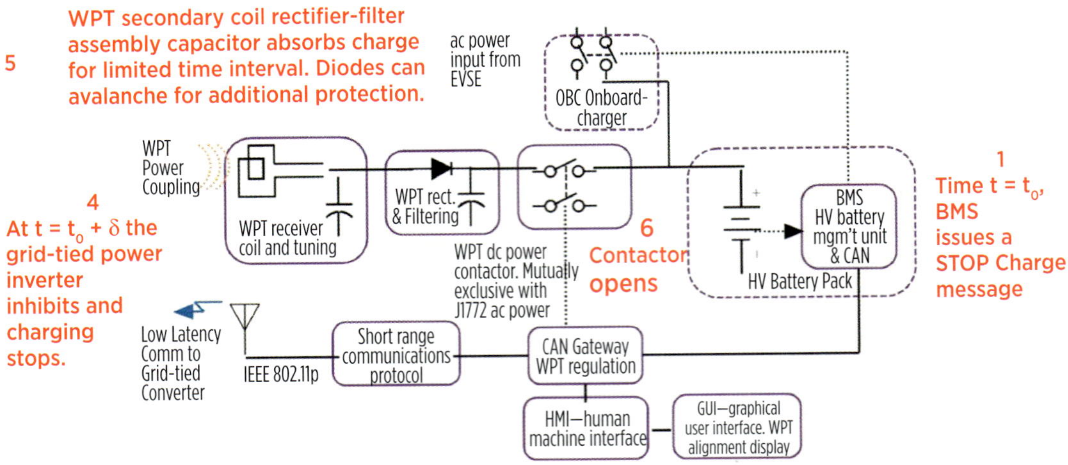


**Figure 5.13** WPT fault detection and second-level response.

The WPT base unit radio modem receives the message to stop charge sent in step 1 and communicates over a local link (i.e., RS232 or RS485) to the AFE and HF inverter controller during step 4. At the WPT base unit, the stop charge command results in disabling of the HF inverter (e.g., inhibit gate drive commands) and disabling of the PFC stage. The PFC stage consists of an inner current regulator that ensures power quality and a voltage outer loop that maintains the HF inverter rail. As soon as the HF inverter is disabled, the rail voltage will rise past its regulation set point, resulting in PFC disabling.

In an emergency shut-down sequence, and due to communications channel latency, one can expect the on-vehicle WPT charge relay to be switched off under load so that some protection must be provided to clamp the charge voltage delivered by the secondary side rectifier (and/or converter if used). This is shown as step 5 in the sequence, during which only the WPT filter capacitor absorbs charge energy. Under high charge rates, the ramp-up time of filter capacitor voltage $U_c$ will be fast (Equation 5.6). This can be seen by noting for a capacitor $C_{filter}$ that $I_{chg}/C_{filter} = U_c/T$. Therefore, some additional protection, such as metal-oxide-varistors (MOV) or other load clamping means, would be provided at the WPT charge relay.

$$T_{limit} = T_{start} + \frac{\left(U_{c\_max} - U_{c\_start}\right)}{I_{chg}} C_{filter} \tag{5.6}$$

For higher-power wireless chargers, the filter capacitance $C_{filter}$ will be larger, but the challenge remains: to prevent overvoltage of the secondary filter capacitor. Furthermore, related WPT components will require a much faster communications channel than DSRC can provide and will use a higher-voltage capacitor and/or additional suppression and fusing.

## 5.7 Normal Shut-Down

The normal shut-down of wireless charging does not pose an emergency condition but can cause problems if not executed properly. Therefore, the concerns expressed for emergency shut-down apply here as do precautions such as limiting and clamping the secondary voltage prior to the charge relay for the high-voltage battery pack. Figure 5.14 schematically shows one such protection scheme in which secondary power after the WPT relay to the vehicle HV battery is switched off such that prolonged exposure is interrupted to over-voltage with an opened fuse element. This, in effect, represents a third level of protection beyond disabling the grid HF inverter and disengagement of the WPT relay. But it results in a case where charging cannot be resumed until a fuse is replaced or a circuit breaker reset. It is not recommended to replace fuses or reset circuit breakers until the cause of the overload condition has been properly investigated and maintenance has been performed.

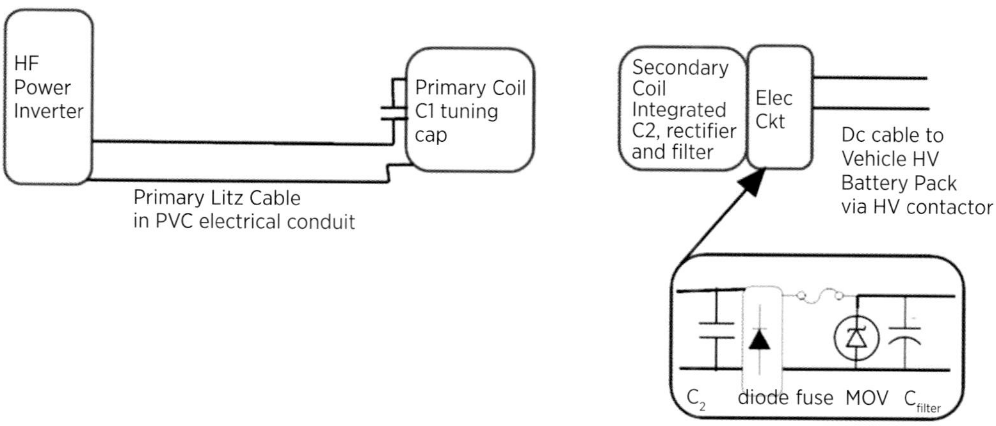


**Figure 5.14** WPT secondary voltage limiting for shut-down protection.

The normal shut-down procedure to terminate charging can be initiated in at least three different ways:

6. The vehicle operator needs to use the vehicle and does a key start.
7. The vehicle battery pack reaches top of charge and sends a terminate charge request over the communications channel.
8. The utility service linked to the V2I radio modem responds to either a TOU or load shedding condition and commands the WPT base unit to terminate charging.

In any of these scenarios, the charging function terminates normally and via the communications channel so that grid-side functions power down first, followed by vehicle-side functions, resulting in the vehicle entering a quiescent, normal key-off state.

Figure 5.14 illustrates the full functional capability of WPT charging for the case when wireless charging power is coordinated with conductive charging. Notice that the vehicle function labeled OBC plus Energy Management Strategy (EMS) plus CAN messaging contains all the features discussed in Figure 5.12 through 5.14 plus existing vehicle ancillary functions necessary to support charging. The highlighted partitions show base and vehicle functionality covered by SAE J2954 wireless charging standard with grid interaction to the left and vehicle interactions to the right. The grid-side shows smart grid interoperability to emphasize that high-power charging of mass market electric vehicles represents a substantial burden and, therefore, must be coordinated. The on-vehicle supporting functions, such as the BMS and the HV to a low-voltage dc-dc converter, are required to be energized during normal operation along with the OBC, or at least some portion of the OBC as needed to fully coordinate wireless and conductive charging operations.

One final note on normal charging relates to the management of "magnetic resonance" in the absence of secondary coil loading. Recall the discussion earlier on the coupling coefficient and what it means for the magnitude of the magnetizing reactance. Without a secondary present, the coupling coefficient is zero, which means the primary inductance is pure leakage. This leakage inductance is tuned out by the primary compensating capacitor shown in Figure 5.15, so the only remaining load on the HF inverter is the primary resistance shown as $R_1$. This would be the HF inverter load, for example, if the vehicle were under change and the operator drove off without a proper shut-down sequence being followed. Referring to Figure 5.15 and assuming for this exercise that the HF source voltage is $U_1(V_{rms})$, the self-resonance of the primary is $\omega_0$, the HF output frequency is $\omega$, the primary current $I_1$ will be the following:

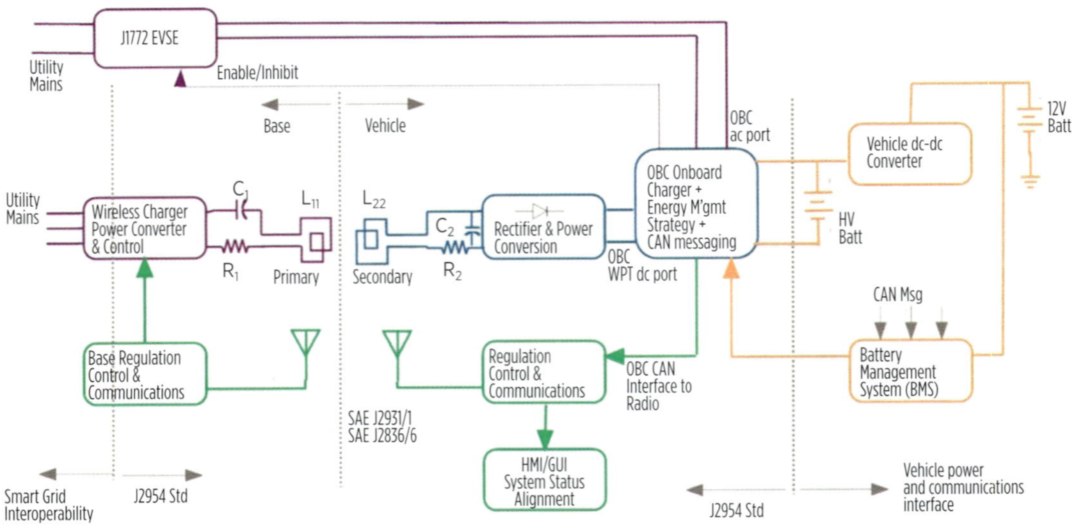


**Figure 5.15** WPT functionality in normal power ON and OFF.

$$I_1 = \frac{\omega C_1 U_1}{\omega C_1 R_1 + j\left(\left(\frac{\omega}{\omega_0}\right)^2 - 1\right)} \overset{\omega \to \omega_0}{\Rightarrow} \frac{U_1}{R_1} \tag{5.7}$$

As indicated in Equation 5.7, a worst-case condition of energizing the primary without a secondary and doing so at the resonant point will result in primary current limited only by the primary cable resistance, which will be very low. The challenge in this case is to manage the HF inverter control so this condition does not occur.

## 5.8 Electrical Safety: High-Frequency Isolation Transformer

Electrical safety is an overarching issue in any type of electrified vehicle charging and, therefore, it must be addressed. In this section, one proposed method of mitigating electrical shock hazard between the wireless charging base unit and the primary pad is the insertion of a high-frequency transformer at the output of the HF inverter H-bridge. One such implementation is shown schematically as Figure 5.16, wherein the transformer isolates the HF cable link to the WPT primary coil, including its integrated tuning capacitor, from the WPT base unit and grid connection.

The safety concern is that any damage to the primary HF cable insulation and conductors that would expose those conductors to being touched would lead to a conductive path back through the WPT base unit and the grid line conductors. The use of a HF isolation transformer addresses this challenge.

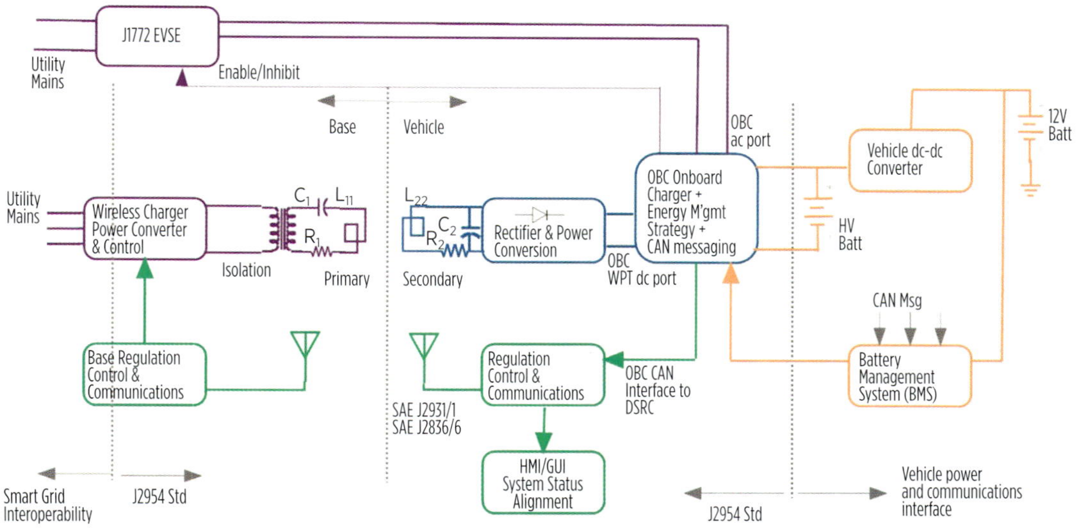


**Figure 5.16** WPT system having a HF isolation transformer at inverter output.

In addition to the galvanic isolation of the primary charging pad from its grid connection, the use of a HF transformer also introduces the flexibility to operate the WPT base unit inverter stage at a higher rail voltage and thereby improve its operating efficiency. A secondary benefit of this flexibility afforded by the transformer turns ratio is that the active front-end power factor corrector stage may be more conveniently fabricated as a boost stage. This benefit is very practical for operation from 200 $V_{ac}$ and higher and for a HF inverter stage operating at bus (i.e., rail) voltages of 340 to 800 $V_{dc}$ or higher.

## 5.9 Emission

In chapter 3, the issue of magnetic field emissions was addressed along with the need to attenuate fringe fields to acceptable levels in the public zone during charging. At present, the recommendations for high-frequency magnetic (and electric) field emissions are to meet ICNIRP1998 levels of 6.25 $\mu$T body average and <87 V/m for the general public. Debate is still ongoing regarding the recommendation of ICNIRP2010 levels that raise this to 27.3 $\mu$T. In all cases, the specifications of emissions are linked to the general public and not to occupational levels that are also reported.

Discussion on where emissions levels are to be measured is ongoing, but most important are all the public zone regions such as at ankle height along the sides of the vehicle under charge and within the passenger compartment on the floor, at occupant seat level, and at headrest level. For WPT installations, the measurements are made primarily at the vehicle perimeter, regardless of where the primary coil is mounted, and in a zone of 0.8 m from the primary charging coil to the measurement point. However, discussion is also addressing having the measurement point at 1.3 m from the primary coil center.

The key point here is that, as with most aspects of wireless charging, recommendations and specifications are challenging and must consider a wide variety of scenarios. Added to this is the variability in emissions due to alignment tolerance discussed in section 5.2.

Electromagnetic field measurements were made on the coupling coils, shown in Figure 5.7 and repeated here, showing the location of a magnetic and electric field measuring instrument, a NARDA EH50D having bandwidth (BW) flat within 0.35 dB from 40 to 100 kHz with sensitivity of 3 nT. Emissions tests were made using the 5 kW rated coupling coils in Figure 5.17 shown at $f$ = 22 kHz, for a working gap $z$ = 150 mm, an inverter duty ratio $d$ = 0.8, and battery voltage $U_b$ = 370 V dc. Locating the NARDA sensor at 0.8 m from the primary coil center is consistent with the nominal track of light-duty passenger vehicles. For example, the Nissan LEAF® EV track (i.e., width between wheels), is 1.53 m with standard P205/55R16 tires, leaving an open width of 1325 m between the tire inner walls and 1.735 m to the tire outside, which is about as close as anyone might stand next to the vehicle. In this case, the distance beneath the vehicle from the center of the primary coil (fully aligned case) to the side of the vehicle would be 0.87 m. Therefore, realistic leakage field measurement points will be in the range of 0.8 m to 0.9 m. Table 5.2 summarizes the magnetic and electric leakage field magnitudes at this point.

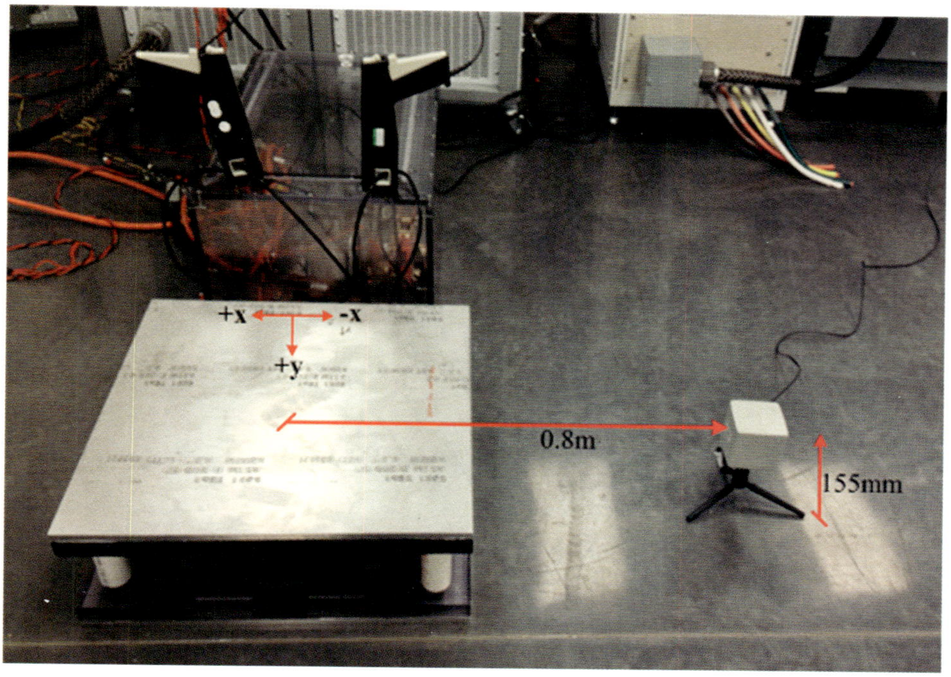


**Figure 5.17** Illustration of leakage field and laboratory measurement location.

**Table 5.2 WPT Leakage Field Levels at 3.3 kW and 5.0 kW Power Level**

| Field component | Battery power = 3.3 kW | Battery power = 5 kW |
|---|---|---|
| *Magnetic field (μT)* | *2.9* | *5.387* |
| *Electric field (V/m)* | *56.6* | *70* |

Additional testing was conducted for leakage fields at higher center frequencies for the WPT coupler shown in Figure 5.17, in this case 22 kHz, 48 kHz, and 80 kHz. The results are plotted against International Standards for Maximum Permissible Exposure (MPE) levels of the magnetic field B ($\mu$T) for each of these frequencies (see Figure 5.18). For this coupler design, the magnetic fringe fields are well within the ICNIRP 1998 level of 6.25 $\mu$T (62.5 mG).

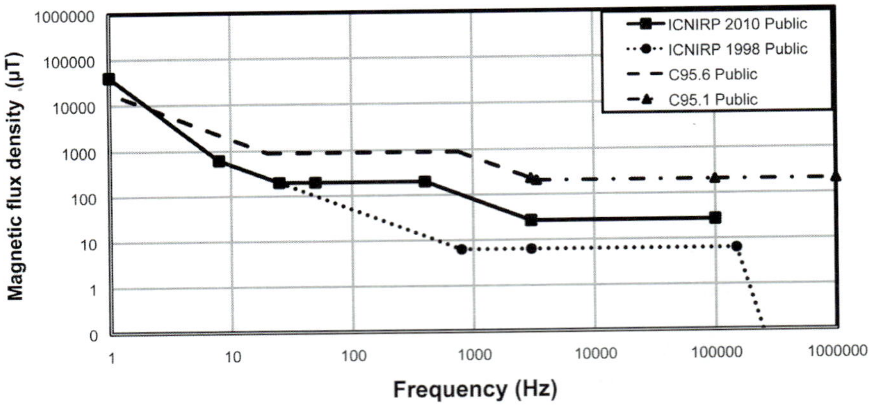


**Figure 5.18** Comparison of WPT B-field levels at 3.3 kW with international standards for maximum permissible exposure levels by ICNIRP and IEEE ([5-6], [5-7], [5-8], [5-9]).

## 5.10 Grid Connection Power Quality

The issue of grid connection power quality is every bit the concern for wireless charging as it is for conductive charging. This means that the same challenges are posed in WPT and conductive charging in terms of meeting THD and the minimum PF. Figure 5.19 is a more comprehensive functional diagram showing the mutually exclusive requirement on conductive and wireless charger options in delivering charge power to the vehicle battery. This mutual exclusivity is also why THD and PF for wireless charging must be on par: both are derived from the same grid and both are delivering charge power to the same battery.

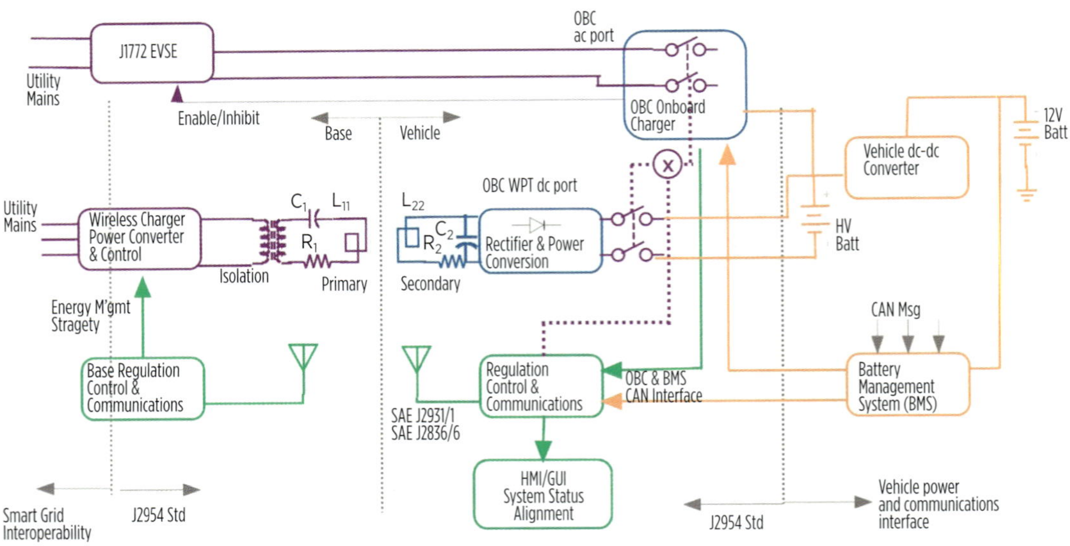


**Figure 5.19** WPT functional exclusivity: wireless or conductive charging.

At the grid connection shown in the upper left of Figure 5.19 for either J1772 EVSE or J2954 WPT charge connection to the utility mains, the expectation is for THD < 5% and PF > 98%. Referring to Figure 5.2, where the active front end is illustrated by a representative PFC stage, the challenge is operating this PFC at sufficiently high frequency so that high-order harmonics are easily filtered using industry standard common mode and differential mode electromagnetic interference (EMI) filtering components. This can be achieved by operating the PFC at a frequency of at least 40 kHz and more typically at 80 kHz and higher. The high-frequency current ripple is then readily filtered and not injected into the grid connection as EMI pollution.

The same can be said for PF. When the PFC is operating and properly synchronized with the utility voltage waveform, the synthesized half sine pulses of current will be within 11.4° of the voltage zero crossings according to Equation 5.8.

$$PF = \cos(\theta) < 0.98 \quad \forall[-11.4° < \theta < -11.4°] \tag{5.8}$$

Figure 5.19 highlights the requirement to inhibit the conductive charger when wireless charging is used and vice-versa. These functions are shown under the control of the communications subsystem regulation and control function discussed earlier in this chapter.

A final comment on power quality concerns for the vehicle HV battery. A conventional OBC unit contains the EMI filtering, PFC, galvanic isolation, and output regulation functions that WPT charging does, but they are not dispersed. All these functions are contained within the integrated OBC and, therefore, are potentially more amenable to control and minimized emissions. Whether integrated, separated, or standalone charger functions, the charge current delivered to the vehicle battery must adhere to the battery and vehicle manufacturer requirements for rate capability versus temperature and SOC as well as maximum ripple current tolerance. Just as with the challenge of THD containment on the grid side, it is recommended that the ripple current to the battery be maintained at less than 15 $A_{pk-pk}$. This will ensure that the charging process will not expose the battery cells to high-frequency ripple currents that may reduce the service life through electrode deterioration (e.g., accelerated aging). This must also be factored into the overall efficiency requirement and be consistent with the point at which WPT efficiency measurements are taken. The recommended location for WPT efficiency monitoring is at the rectified and filtered output of the vehicle mounted secondary, shown as the function block connecting to the WPT HV contactor. This means that, for a contactor mounted with the secondary coil and electronics, neither the cabling route through the vehicle to the battery tray nor the main system relay, if contained within the battery tray, are being counted in the overall efficiency. These are ancillary issues to be addressed by field measurement and validation testing as well as standards validation groups.

## 5.11 Installation and Commissioning

Just as with conductive charging installations, such as SAE J1772 EVSE, the expectation is that WPT charging installations will be subject to the same, if not identical, procedures. For example, a residential customer wishing to wirelessly charge a PEV or AEV will encounter the following:

- Notification to the utility provider that a charging station is planned and its charge rating
- Procurement of a WPT base unit from a certified manufacturer that is known to be compatible with the customer's PEV or AEV equipped with an OEM-installed wireless charger option
- Contracting with a certified electrical contractor to install the electrical service from the utility service entrance via an approved meter (that in a smart grid may be equipped with advanced metering infrastructure, AMI) to the customer's charging site
- Approval of the WPT base unit installation by the municipal electrical inspector or other AHJ
- Field measurement and validation by a certified installer, who could be the electrical contractor, or the municipal or AHJ inspector, using a standard secondary coil and monitoring equipment to verify that the installation meets applicable regulations and standards

The last two points in the list may be interchanged depending on how WPT standardization activities progress and are adopted. At this time, a determination on field measurement and validation has not been made, but in the United States, it is anticipated that Underwriters Laboratory (UL) certification will be mandatory for the installation to meet fire protection agency requirements and, hence, the home insurer requirements for adequacy of coverage.

## 5.12 Summary

This chapter presented a wide variety of technical issues of wireless charging technology ranging from safety aspects of WPT chargers to its attendant communications needs and challenges, through electromagnetic field emissions monitoring and containment, and on to field installation, measurement, validation, and commissioning. Wireless charging of PEVs and AEVs has the potential to eclipse conductive chargers in convenience, flexibility, and future enhancements to fast charge and eventually in-motion charging. These advancements hinge on successful development and deployment of stationary charging in the mass market before moving to heavy-duty vehicles such as shuttle and transit buses and then rail and other forms of mass transit. The issues and challenges in the many aspects of WPT deployment have been discussed, and the stage has been set for adoption in the mass market.

## References


5-1. G. A. Covic, J. T. Boys, A. M. W. Tam, and J. C-H. Peng, "Self Tuning Pick-ups for Inductive Power Transfer," IEEE Power Electronics Specialists Conference, PESC2008.

5-2. F. Musavi, W. Eberle, and W. G. Dunford, "A High-Performance Single-Phase Bridgeless Interleaved PFC Converter for Plug-in Hybrid Electric Vehicle Battery Chargers," IEEE Transactions on Industry Applications, vol. 47, no. 4, pp.1833–1843, July–Aug. 2011.

5-3. L. Huber, Y. T. Jang, and M. M. Jovanovic, "Performance Evaluation of Bridgeless PFC Boost Rectifiers," IEEE Transactions on Power Electronics, vol. 23, no. 3, pp. 1381, 1390, May 2008 (doi: 10.1109/TPEL.2008.921107).

5-4. J. M. Miller, O. C. Onar, C. P. White, and P. W. Ryan, "Grid Side Regulation of Wireless Power Charging of Plug-in Electric Vehicles," IEEE 4th Energy Conversion Congress & Exposition, ECCE2012, Raleigh, NC, Sept. 2012.

5-5. J. M. Miller, "Wireless Power Charging Fundamentals and Challenges," SAE International 2012 Electric Vehicle Technologies Symposium, 23 February 2012, San Diego, CA.

5-6. Guidelines for Limiting Exposure to Time-Varying Electric, Magnetic, and Electromagnetic Fields (Up to 300 GHz), ICNIRP Guidelines, Health Physics Society, 1998.

5-7. ICNIRP GUIDELINES—For Limiting Exposure to Time-varying Electric and Magnetic Fields (1 Hz–100 kHz), HEALTH PHYSICS 99(6):818–836; 2010.

5-8. IEEE Standard for Safety Levels with Respect to Human Exposure to Radio Frequency Electromagnetic Fields, 3 kHz to 300 GHz, IEEE Standard C95.1, 2005.

5-9. IEEE Standard for Safety Levels with Respect to Human Exposure to Electromagnetic Fields, 0–3 kHz, IEEE Standard C95.6, 2002.

# Chapter 6
## Markets, Strategies, and Standards for EVs with WPT

### 6.1 Introduction

Constantly rising fuel prices encourage the automotive industry to develop electrified powertrains for alternative vehicle concepts. Since WPT systems radically simplify the charging procedure, they are considered one of the main market drivers for electric vehicles. For example, customers of today's conventional plug-in hybrids don't always charge their batteries via cable, but rather go to the next gas station. Hybrid vehicles with WPT systems are more likely to be charged. Automated, more convenient, and thus more frequent wireless charging reduces the depth of discharge of batteries and can potentially extend battery life and increase the electrical driving-mode portion of hybrid vehicles [6-1]. WPT systems have been developed to reap these benefits, which have been evaluated by an increasing number of test vehicles—which have different powertrain concepts such as PHEV, range-extended electric vehicle (REEV), and BEV—and test trials over the past few years. However, some major obstacles still remain to be overcome on the way to a profound global market introduction of WPT systems, including safety, standardization, and cost reduction.

Several standardization committees, such as the DKE VDE-AR-E 2122-4-2, 352.0.1, the IEC 61980, ISO 19363, and the SAE J2954 are working with industry to consider and coordinate a variety of different interests in the topic from the vehicle and the infrastructure side. On one side, energy and infrastructure providers see great potential for wireless charging and automated coupling and thus, higher availability of battery storage capacities of vehicle fleets for the energy grid. On the other side, vehicle companies want to ensure control over their vehicles and a lifetime-optimized operation of vehicle components, especially the batteries. The standardization process is complicated further by an automotive market in which car manufacturers vie to be the first to introduce innovative product features. However, car manufacturers are aware that a standardized charging infrastructure with interoperability between the different charging systems is a basic requirement for further market development of WPT systems.

Besides standardization, cost reduction of WPT systems is also vital to compete with conductive charging solutions. The basic structure of WPT systems for electric vehicles consists of roadside energy-supply components and coil structures (primary coil), and vehicle-side power pick-up (secondary coil), which includes rectification and appropriate communication and surveillance systems [6-2]. The market price of wireless power transfer components will depend on the specifications required by car manufacturers (OEMs) and the first-tier suppliers. A competitive supplier and OEM market will ensure a fast reduction of the market price so that WPT systems can also be offered for mass market vehicles in cheaper price segments. Therefore, design-to-cost and intelligent production approaches are necessary to develop and produce the charging equipment at the lowest possible cost while ensuring safe and reliable integration into vehicle concepts. So far, WPT systems for electric vehicles have been produced mainly in manual prototypical production processes in small numbers. However, it is important that product quality can be ensured by well-established production processes. Thus, a particular challenge for prospective mass production is to deliver, handle, and process sensitive materials reliably in semi- and high-automated production processes [6-3]. Therefore, flexible robot-based assembly and handling technologies are needed [6-4], [6-5], as shown for Litz wire coil packages in Figure 6.1.

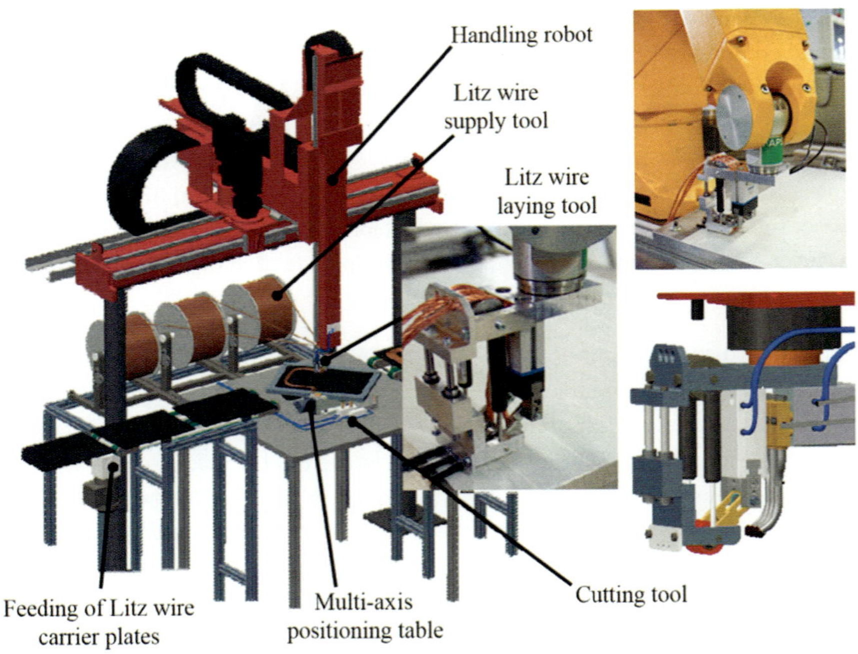


**Figure 6.1** Flexible robot-based assembly cell.

Aside from stationary WPT charging options, vehicles can also be powered with WPT systems while in motion. This is of specific importance for fields of application in which a conductive plug-in system is impracticable. This is the case at taxi stands where taxis move slowly to the front of the line while waiting for customers. Charging of each single primary pad underneath the road surface of the taxi stand can be activated once the receiving pads of the taxis are aligned above them. Such a semi-dynamic charging could also be implemented in urban areas at traffic lights or intersections where vehicles are moving slowly, or stationary for a certain amount of time.

Another important marketing step could be dynamic charging, which happens in motion, even at higher speeds, through WPT systems that are embedded in the roads. The visionary idea of picking up energy while driving, which would eliminate the driver's range anxiety regarding battery life, has been discussed widely in recent years. In theory, it is a promising approach to keep the energy in the grid, away from expensive and heavy batteries, making vehicles cheaper and lighter. Therefore, several test vehicles have been set up. For example, the company Qualcomm HaloIPT is working with the FIA Formula E Team of Drayson Racing to evaluate a specially developed 20 kW high-power, stationary wireless charging system in prototypical electric racing cars [6-6]. First, the stationary WPT system will simplify charging and improve safety as it eliminates loose cables and exposed connectors within the pits. In the future, dynamic wireless charging will be investigated to charge vehicles during the race on the track to reduce or even eliminate the need to stop the cars for recharging. Such motorsport approaches offer the potential to develop systems that are much more sophisticated in terms of lightweight design, system integration into the chassis, and new cooling concepts for high-power systems, and they will rapidly push mechanical and materials development, resulting in lighter, more efficient, and more reliable systems [6-6].

However, it is not yet possible to derive a functional business model and stakeholders for a public dynamic charging infrastructure for private passenger cars. Private car trips are usually not very long, and the average car is parked most of the time either at home or at work. In Germany, 43% of private car routes are recurrent. The typical daily travel time for private cars is 32 min, with an average driving time of 21 min and covered distance of 15 km (9.3 mi). The majority of private trips are shorter than 40 km (24.9 mi) [6-7]. For electrified vehicle charging concepts, stationary WPT charging stations at home or at work are the most promising and sufficient for a market introduction. A public charging infrastructure, whether stationary or dynamic, is expensive, and it is difficult to win support from policy-makers and stakeholders, especially due to the variability of private car driving routes [6-1]. Dynamic charging approaches for passenger cars will need government initiatives to ensure that the right business models are possible for industry to invest in the infrastructure.

WPT technology also proves to be increasingly applicable and technically mature for fleet operators in the field of freight and passenger logistics, with power transfer rates in the range of up to several hundred kilowatts. Different system suppliers have successfully tested and demonstrated these high power rates for trams and buses [6-8], [6-9].

First market scenarios can be found, especially for such commercially operated fleet vehicles, because their definable, limited operating environments allow for a precise cost calculation. Therefore, different fields of application for WPT systems are classified in Table 6.1 according to the needed operating range and presence of predefined routes.

**Table 6.1 Classification of Fields of Application for WPT Systems [6-5]**

| | Small operating range/ trips with intermittent stops (< 5 km radius) (< 3 mi radius) | Medium operating range/trips with fewer intermittent stops (5–50 km radius) (3–30 mi radius) | Large operating range/ trips with almost no intermittent stops (> 50 km radius) (> 30 mi radius) |
|---|---|---|---|
| Predefined fixed routes | Theme park routes/trolley trams/AGV intra-logistics/ electric motor racing | Local public transit, garbage disposal service | Fixed long-distance heavy truck routes |
| Occasionally varying routes | Intra-logistics (forklift, reach truck) | Commuter, city logistics | Special delivery long-distance freight operations |
| Always varying routes | AGV harbor logistics/ airport logistics | Car-sharing, taxis | Business representatives |

Planning a WPT infrastructure becomes increasingly difficult when the routes are less predefined and the operating range is larger. Therefore, the following chapter presents the optimization problem for WPT in electrified vehicles and a simulation approach with a discrete-event simulation tool to determine an optimized WPT infrastructure design. Subsequently, three different fleet-operated applications from Table 6.1 are selected, described, and simulated with respect to their suitability to find an optimized WPT infrastructure allocation at low costs: automated guided vehicles (AGVs) in harbor logistics (small operating range), local public transport (medium operating range), and long-distance heavy truck transport (large operating range).

## 6.2 Optimization Problem for WPT in Electrified Vehicles

The installation of a WPT infrastructure is mainly determined by battery size in the vehicles, the total length of the installed power transfer segments, and the vehicle's power transfer rate. The goal of the optimization is to balance battery size with the allocated infrastructure to eventually meet range requirements of electrified vehicles at the lowest costs [6-5], [6-10].

### 6.2.1 Battery Size (kWh)

More charging units with high power transfer rates along the road allow for more frequent charging and reduce the needed battery size. As battery capacity and weight of electric vehicles decrease, they require less power, which additionally reduces vehicle cost. However, lower battery capacity also leads to range reduction and higher dependence on the infrastructure.

### 6.2.2 Power Transfer Rate (kW)

The higher the power transfer rate, the fewer meters of infrastructure segments are required, since more power can be provided over a shorter length. A high transfer rate also reduces battery capacity needs because the batteries can be charged faster (in case of a fast chargeable battery system). However, high-power components, which are required for the generation and storage of the high power transfer rates, cause higher costs.

### 6.2.3 Infrastructure Allocation

A generous distribution of power transfer infrastructures allows for more frequent power transfer to the vehicles, and thus, a reduction of battery capacity and/or the power transfer rate. An excessive allocation of charging infrastructures, however, comes with high installation costs. The right allocation of power transfer segments containing stationary and/or in-motion power transfer facilities along the routes is of great importance for overall cost and varies depending on the application and charging strategy of the infrastructure and vehicle operators.

To reduce the complexity of the optimization problem, general rules for the allocation of the infrastructure need to be developed. These rules help determine an appropriate tradeoff between infrastructure investment and necessary battery size. They enable stable system behavior with low operating and installation costs. Especially for various fleet-operated vehicle scenarios, clearly defined and structured allocation rules for the installation of a charging infrastructure can be derived from an energy ("help with energy") and power ("help with power") perspective, as shown Figure 6.2.

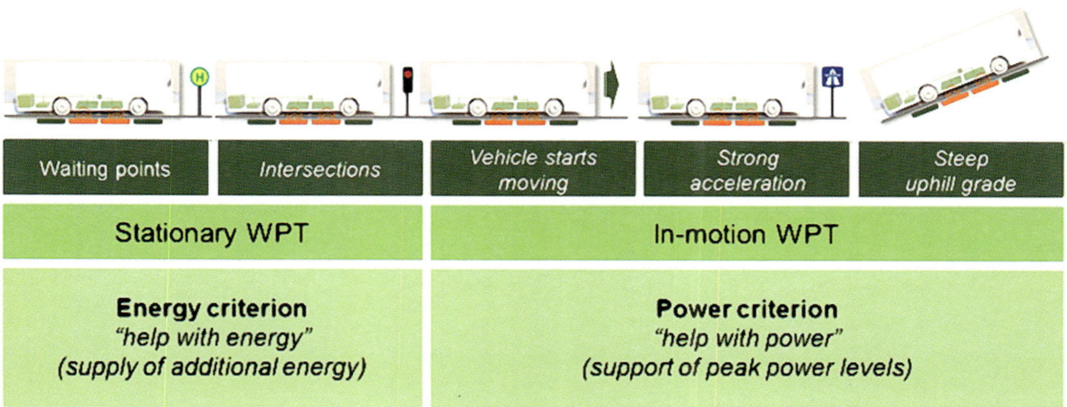


**Figure 6.2** Allocation rules for the installation of a WPT infrastructure [6-5].

The power transfer infrastructure can be placed at stops, intersections, or even at specific points along the route for stationary power transfer ("help with energy"), raising the SOC of the battery. Therefore, the electric motor(s) can be operated from the energy stored in the battery when no infrastructure is available. When the vehicle is stationary or driving on a route where the transmitters are installed, the battery can be charged, and the vehicle drive system and/or vehicle auxiliary loads (e.g., air conditioning unit) can be powered directly ("help with power"). For peak power levels that occur during strong acceleration or on steep uphill slopes, strong charging and discharging operations may rapidly deteriorate even modern lithium-ion batteries. Thus, this is important to extend the vehicle's battery lifetime.

These general allocation rules help to reduce the planning complexity by ensuring that enough power can be transferred to the vehicles under all given circumstances. However, the planning complexity increases dramatically when the vehicles have no strict track guidance, as is the case for most electric vehicle applications. The more complex and less predefined the routes are, the more assumptions have to be made and the more complex algorithms are necessary for the respective planning approaches. The following section presents a simulation-based planning approach that aims to solve the mentioned optimization problem even for complex WPT traffic scenarios.

## 6.3 Simulation Approach to Determine an Optimized WPT Infrastructure Design

Reducing uncertainty in the planning process will encourage future acceptance and spread of wireless powered vehicle concepts. This will require comprehensible and user-friendly planning and decision-support tools to determine the appropriate tradeoff between infrastructure investment and needed battery size for stable system behavior, at low cost, and demonstrating the technical and economic appeal of WPT in electric vehicles. Therefore, this chapter shows a new approach to a profound decision-making basis with the support of a discrete-event simulation tool, extended by detailed energy input/output relations. This approach enables potential customers of WPT systems to calculate individual scenarios quickly and to form the basis for substantiated technical and economic feasibility studies [6-5]. Figure 6.3 shows the input and output factors of the optimization problem and their relations in an energy model.

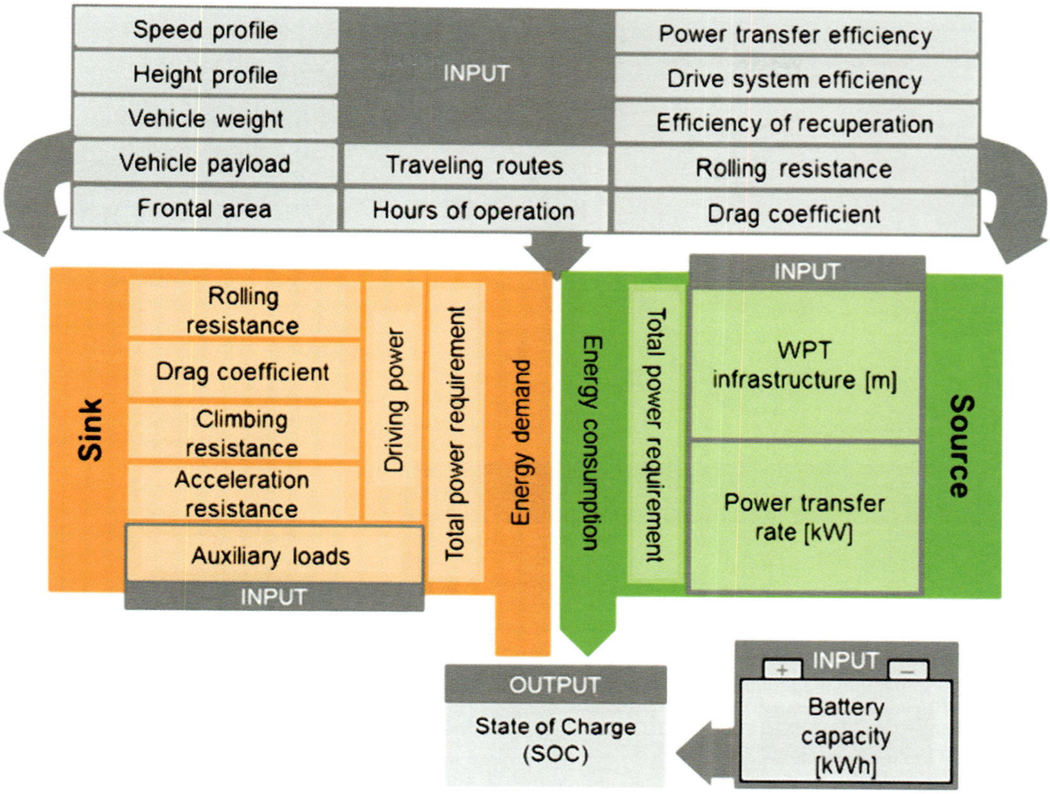


**Figure 6.3** Energy model for the simulation of the SOC.

To allocate WPT infrastructures economically and to evaluate the proper power transfer rates and vehicle battery capacities, the Institute for Factory Automation and Production Systems (FAPS) of the Friedrich-Alexander-University Erlangen-Nuremberg (FAU) is using the discrete event simulation software Plant Simulation from Tecnomatix™. To meet the demands of a WPT infrastructure design, the software has been adapted and expanded with the necessary energy parameters and traffic flow logic (Figure 6.4).

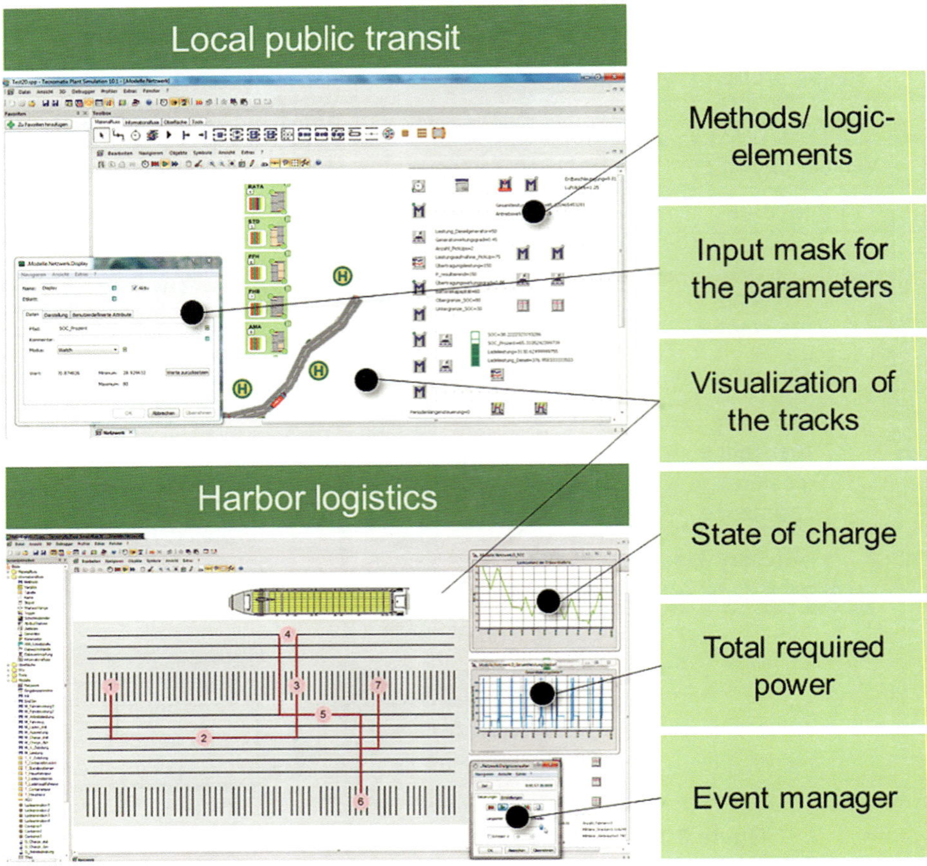


**Figure 6.4** Plant simulation software to calculate WPT traffic scenarios (© 2012 IEEE. Reprinted, with permission, from [6-3]).

Plant Simulation is a module of the Tecnomatix™ solution suite and is already being used successfully for planning and programming of complex production processes and logistics operations [6-11]. The powerful simulator with its open system architecture allows for the creation of individual customized attributes with good graphic display options. In addition, Plant Simulation has powerful analysis tools, including automated bottleneck analysis, Sankey diagrams and Gantt charts, and a simple, easy-to-use interface. Furthermore, the standard vehicles in the software, such as the ones used for logistics operations, can be adapted for WPT operation. The object-oriented hierarchical simulator also allows for modeling complex drive-control strategies and for running experiments. With an average training and adaption time, it is possible to model, simulate, analyze, and lastly optimize fleet-operated electric vehicle scenarios in a reasonable amount of time, despite the fact that Plant Simulation is a production and logistics flow simulation software and has not been used for traffic scenarios so far. The next chapter derives the potential infrastructure parameters, technical feasibility, and economic implications based on Plant Simulation for three specific fields of application of high-performance transportation (harbor logistics, local public transport, long-distance heavy-truck transport).

## 6.4 Fields of Application for Fleet-Operated Vehicles Using WPT Systems

### 6.4.1 Harbor Logistics (Small Operating Range)

Automated guided vehicles (AGVs) are used at harbor container terminals for handling containers between the quay and storage areas. Downtimes caused by currently used battery exchange systems [6-12] can be eliminated using WPT systems, enabling a continuous 24-h operation of zero-emission AGVs. The routes of the AGVs between the quay and the storage area are not unique, but they show a recurring behavior. All the routes have the same cornerstones (waiting positions, loading positions, main traveling lanes). Therefore, the driving route of an AGV always includes the following: waiting for a new job in a stationary position, driving toward the container bridge, standing in a waiting position once again to hold the correct sequence, loading containers from the container bridge, driving to a storage area, unloading containers at the storage area, and driving to a new waiting position until the next order. Thus, starting and destination points are clearly defined. However, the selected path to get there is variable. The average travel time of the AGVs in a 24-h operation at a given container terminal is four minutes. The waiting time is 50% of the total time, and the average path length of one order is 460 m [6-12]. In principle, both stationary and in-motion WPT are conceivable for AGV harbor logistics operations. Waiting, container bridge, and storage area positions are potential stationary power transfer points. Thus, the AGV battery can be charged while waiting for, picking up, and unloading containers. Also, in-motion WPT is possible along the main traveling lanes, with AGV batteries being charged as they drive between the different target positions ("help with energy"), or the powertrain being directly powered from the power transfer segments ("help with power"). Several possibilities for infrastructure allocation exist. A simulation of a sample container terminal with AGVs shows that with high-power systems, even a partial allocation with WPT at the waiting positions is sufficient to power the vehicles, as shown in Figure 6.5.

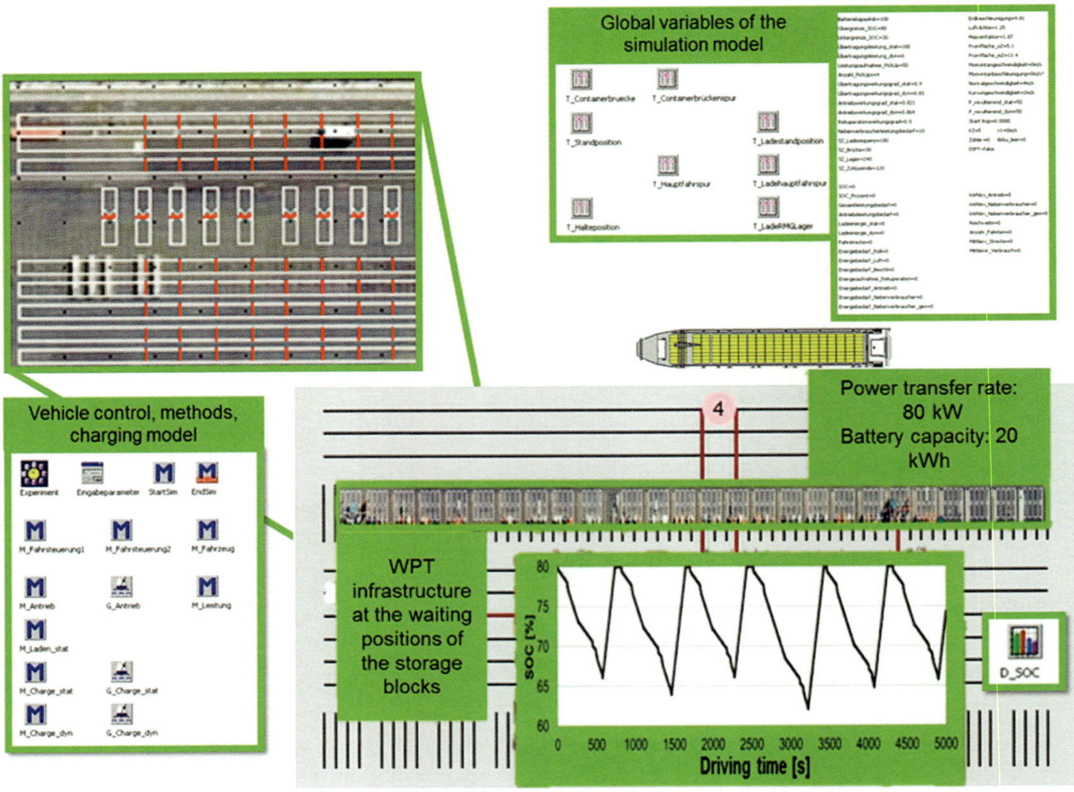


**Figure 6.5** Optimized WPT infrastructure equipment for AGVs.

In this case study, the waiting positions at the storage blocks should be equipped with stationary power segments and a power transfer rate of at least 80 kW (Figure 6.6, number 2). The WPT-AGVs have to be equipped with a fast chargeable battery capacity of at least 20 kWh (Figure 6.6, number 1). Overall, the infrastructure must provide a charging energy of about 35 MWh per day to guarantee a smooth operation. Thus, it could be shown that an area-wide allocation in closed systems, such as a container terminal in a harbor environment, allows for a charging strategy that only maintains the charge level from power transfer segment to power transfer segment, with a minimal battery capacity (Figure 6.6, number 1) and a minimized power transfer rate (Figure 6.6, number 2).

To reduce power transfer segment coverage further, the charging strategy must focus on significantly larger power transfer rates and larger battery capacities to avoid completely depleting the AGVs' batteries between the power transfer segments.

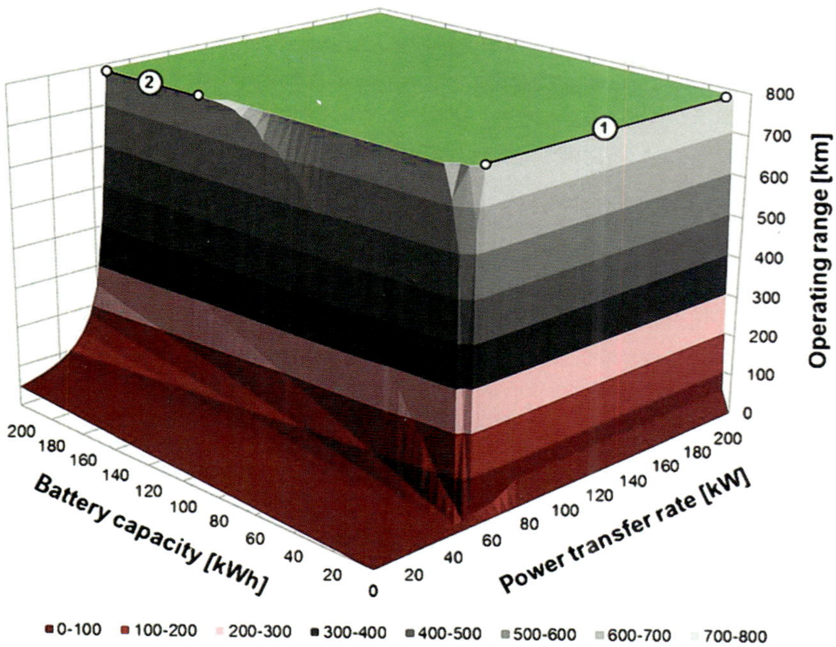


**Figure 6.6** Range in km attained with WPT equipment if offered at all waiting positions (100% coverage) [6-5].

More simulation models and especially real-world high power WPT demonstrations have to be set up to verify the technical input parameters of the WPT systems. The first results of the simulation tool already show the promising technical feasibility of WPT systems for harbor port logistics, if the technical parameters of initial demonstration projects can also be achieved in a harbor logistics environment. The total cost over the lifetime of the WPT systems compared to conventional diesel-operated vehicles depends on the evolution of energy prices, the costs for smaller but more expensive fast chargeable battery systems, and above all, on the infrastructure and maintenance costs that can ultimately be attained. Today, a WPT infrastructure for dynamic charging is still expensive, especially during the initial market introduction phase and due to safety concerns about the integration of high-voltage cables in the ground. It must also be considered that the ground settlement behavior at newly deposited harbor quays may cause severe problems, as there are strict requirements to promptly renew road surfaces at regular intervals. This makes stationary power transfer solutions more likely but still demanding in harbor environments.

## 6.4.2 Local Public Transport (Medium Operating Range)

WPT technology offers the chance to establish emission-free local public transport at lower battery costs. Today, electric buses or trams are generally powered via overhead contact lines. Overhead contact lines are exposed to the weather and, therefore, incur significant maintenance costs. A roadway-embedded WPT infrastructure can solve these issues and additionally preserve the aesthetics of historic cityscapes [6-9]. City trams and buses usually move in fixed, predefined travel routes. A sample bus route is about 13 km (8.1 mi) long, travel time is 55 min, and 11 buses are used for 10 min intervals between buses. During their average of 14.5 h of operation per day, buses are stationary for about 35% of the time. Stations for loading/unloading passengers, and the speed of the vehicle at specific locations on the route are predefined. Thus, WPT can be applied at intermediate stops while passengers are boarding and exiting the bus and/or at final stops with longer-than-average stopping times. Road conditions such as uphill or downhill slopes can also easily be determined, making it easy to assess power requirements for the vehicles at specific points. In addition, in-motion power transfer or direct powering of the powertrain while driving is possible. This allows for smaller batteries or, in hybrid buses, a larger portion of pure electric drive mode.

The simulations conducted with Plant Simulation show the technical feasibility for future local public transport applications of WPT technology. With a sufficient battery capacity and enough WPT segments at the terminal and at bus stops, it is possible to operate even purely electrically powered buses. Consequently, for some bus routes, and depending on energy price scenarios, WPT technology has the potential to electrify bus operation and thus to reduce operating costs of local public transit authorities. Based on a selected, representative bus route, it could be shown that it can be operated by equipping the two terminal bus stations with 200 kW power transfer systems, and six other bus stops in-between with 100 kW (Figure 6.7; Figure 6.8, number 1).

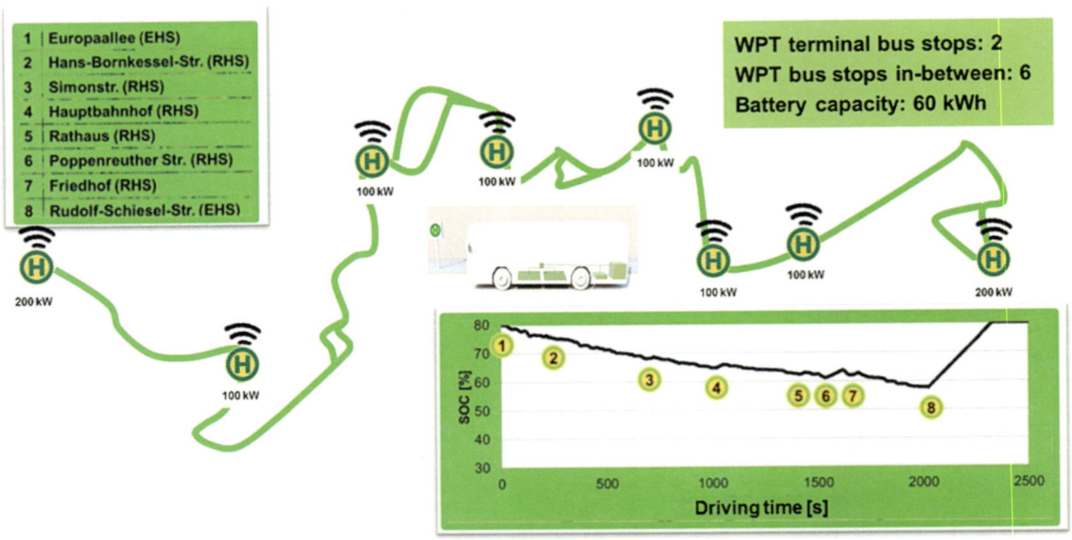


**Figure 6.7** Optimized WPT infrastructure design for a representative bus line.

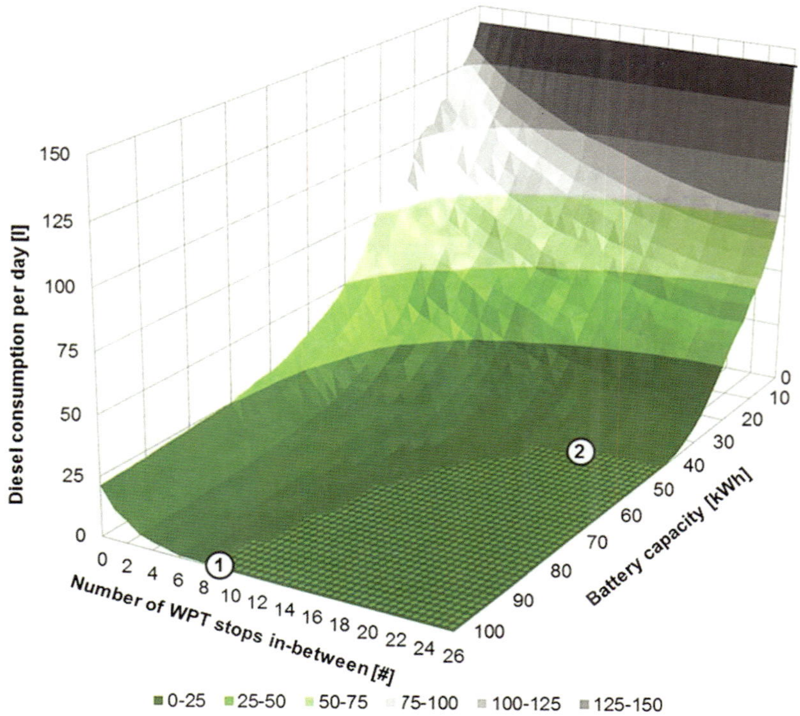


**Figure 6.8** Diesel consumption per day for a WPT hybrid bus in correlation of different battery capacities and infrastructure allocation scenarios, based on the examined bus line.

It could also be shown that the buses in this study could be operated purely electrically. However, the WPT bus has to be equipped with an expensive fast-chargeable battery system with a battery capacity of at least 50 kWh (Figure 6.8, number 2).

As a result, it may be preferable to use a WPT hybrid bus concept to deal with any faults and shortcomings in the initial phase of a market introduction to avoid any service disruption. In the simulation run, 73% of the daily energy demand of the WPT hybrid bus of 574 kW was covered at the terminal bus stops, 24% at the bus stops in-between, and only about 3% by the diesel engine. In conclusion, it can be said that WPT systems for public transport are technically feasible. They will become increasingly economical in the future as diesel prices rise and the costs for infrastructure and for batteries decrease.

### 6.4.3 Long-Distance Heavy Truck Transport (Large Operating Range)

The major part of freight transport volume in Germany is operated on roads by heavy trucks with a maximum permissible weight of 40 t [6-13]. The option of electric trucks is appealing, as electric engines and longer service lives of the drivetrains increase long-term economic efficiency [6-14]. So far, heavy and expensive batteries have been crucial barriers to the electrification of heavy trucks. On the one hand, the travel routes of long-distance freight trucks for standard orders of shipping companies run on predefined routes. On the other hand, the travel routes strongly vary for special-delivery orders. Most of the trips take place on highways at higher speed. Thus, trucks in long-distance operations need to be powered in motion directly while driving on the highway. Without directly powering the drive systems, the large demand for power cannot be met, especially not on steep uphill slopes. The necessary batteries would be so heavy that almost no carrying capacity would be left. A continuous power transfer to trucks while in motion could reduce or even totally eliminate this restriction, depending on the extent of installed infrastructure. Thus, significantly smaller and lighter batteries would be sufficient, enabling a total or at least partial electrification of long-distance heavy-truck traffic. This would require electrified roads powering the trucks in motion along the right lanes of highways ("help with power"). Additionally, stationary power transfer points at parking lots along the highways could be used to charge the batteries during mandatory truck driver rest periods ("help with energy"). Long-distance freight transport could be considered the most demanding application for WPT. However, if it can be achieved under economically reasonable conditions, it has enormous potential both for the national economy in terms of job creation and environmental protection in terms of reducing concentrated carbon dioxide emissions [6-5], [6-14].

To evaluate the appeal of in-motion, long-distance, wirelessly powered heavy truck transport, an uphill highway section (Kindinger Berg, Bavaria, Germany) has been selected for the simulation. Due to the steep uphill slope, the Kindinger Berg has significantly high energy requirements for truck traffic. The maximum power requirement of a fully loaded truck (40 t) to pass the Kindinger Berg is about 390 kW, and for one passage, each heavy-duty truck needs about 27 kWh. To pass the Kindinger Berg, a power transfer rate of 40 kW per pick-up is necessary and technically feasible. Thus, in this simulation scenario, the trucks have to be equipped with 13 pick-ups at the underbody of the vehicle so that even fully loaded heavy trucks (40 t) can pass the Kindinger Berg with a maximum speed of 45 km/h (28 mph).

The entire infrastructure section reaches its maximum power of about 30 MW when several trucks are crossing the Kindinger Berg at maximum speed and minimum safety distance between trucks. The equipment of the Kindinger Berg with WPT segments helps WPT hybrid heavy-duty trucks to pass this uphill highway section using purely electric power. However, it is not profitable for shipping companies to equip only very few highway sections. If multiple energy-intensive sections of highways are equipped with WPT, the trucks can use them for a significant portion of their travel routes. The conducted simulations show that the use of in-motion WPT for long-distance heavy trucks seems to be visionary but technically feasible if the technical parameters of the first demonstration projects can be achieved. Furthermore, it could be economically quite advantageous in a future context of rising fuel prices, and of significant synergies with a smart energy grid based in large part on renewable energy sources [6-15]. The traffic and energy grid could be better balanced with electrified truck traffic that is shifted to off-peak hours with the help of toll rates that vary according to the time of day.

Aside from promising results with higher power rates with WPT techniques, project approaches with overhead contact lines for trucks have also been started and have intensified the debate about the demanding and visionary option of long-distance electric trucks [6-16]. The multi-use capabilities of wireless electrified roads for other road users and the positive effects on aesthetic landscapes speak to the value of WPT systems. However, installation and maintenance costs for all these systems, especially WPT systems, have to be reduced by new roadway construction methods [6-17]. These costs are difficult to derive; therefore, further test tracks have to be set up at different research locations [6-8], [6-9], [6-14].

## 6.5 Conclusions

In view of rising global interest and first announcements of the launch of stationary WPT systems for passenger cars [6-18], [6-19], [6-20], it is clear that considerable new opportunities for value creation have opened up, especially in combination with a general increase in the demand for electric vehicles. However, issues such as standardization and safety still have to be carefully addressed, and component costs must be reduced to compete with conductive charging solutions in the foreseeable future.

Public stationary and especially dynamic wireless charging solutions pose much greater challenges, especially in terms of proving their technical and economic efficiency. It can be expected that initially, smaller WPT systems with clearly predefined routes will be installed for application. The technology makes it possible, for example, to bypass infrastructure installations (e.g., at complex intersections) and preserve aesthetic cityscapes, but it also competes with other short-term supply options (e.g., energy storage in the form of a smaller battery or supercapacitor that can also be charged via a pantograph). Whether WPT systems can prevail as public charging solutions for electrified vehicle concepts finally depends on the attitudes and responses of local authorities, especially those in charge of roads and transportation. Therefore, WPT technology providers have to support potential customers with planning tools to convince them of the advantages of this technology.

Using the simulation models shown in this chapter, the optimal combination of battery capacity (kWh), power transfer rate (kW), and charging infrastructure allocation (m) can be reliably determined. This helps to design technically and economically optimized WPT systems early in the planning phase. The optimization task for WPT infrastructure design can be solved with even less effort when the number of variables can be limited in advance by setting general rules (e.g., the presented allocation rules). These rules are usually easy to derive from the analyzed system by examining the wait times of the model vehicle on the infrastructure and by identifying which parts of the travel routes pose increased power requirements. However, the output of these models can only be as accurate as the input variables. As deterministic assumptions and given settings may not perfectly reflect the characteristics of an actual vehicle operation under all possible circumstances (e.g., cold weather, mass events in a city, traffic jams), a specific tolerance factor has to be factored in to guarantee reliable operation and to supply enough power under almost any conceivable scenario. It is vital to critically review results and assumptions. Detailed cost data must be collected for each scenario in future demonstration projects, as they always depend on local circumstances such as grid connection, ground stability, and the time available for the construction and maintenance of the infrastructure. As soon as detailed cost data are available, cost simulations can easily be added to the presented simulation tools.

## References


6-1. F. Risch, "Application Study of Wireless Electric Vehicles," Pre-Conference Tutorial, Proceedings of Int. Conference on Electric Roads & Vehicles (CERV): Park City, Utah, 2012.

6-2. M. Budhia, G. A. Covic, and J. T. Boys, "A new IPT magnetic coupler for electric vehicle charging systems," 36th Annual Conference of the IEEE Industrial Electronics Society, IECON: Phoenix, Arizona: pp. 2487–2492, 2010.

6-3. F. Risch, S. Guenther, and J. Franke, "Production Concepts for Inductive Power Transfer Systems for Electric Vehicles," Proceedings of the 1st Int. Energy Transfer for Electric Vehicles Conference (E|TEV): Nuremberg, Germany, 2012.

6-4. F. Risch, S. Guenther, B. Bickel, and J. Franke, "Flexible Automation for the Production of Contactless Power Transfer Systems for Electric Vehicles," Proceedings of the 2nd Int. Energy Transfer for Electric Vehicles Conference (E|TEV): Nuremberg, Germany, 2013.

6-5. F. Risch, "Planning and Production Concepts for Contactless Power Transfer Systems for Electric Vehicles," Institute for Factory Automation and Production Systems, Friedrich-Alexander-University, Erlangen-Nuremberg, PhD-thesis, to be published, 2014.

6-6. G. Ombach, "Wireless EV Charging, Validation and Compliance for Public Trials," Proceedings of the 1st Int. Energy Transfer for Electric Vehicles Conference (E|TEV): Nuremberg, Germany, 2012.

6-7. n.a., "Mobilitaet in Deutschland 2008," Federal Ministry of Transport, Building and Urban Development: Bonn, Germany, 2009.

6-8. I.-S. Suh, "Innovative EV Charging with OLEV Technology and its Practical Applicability," Proceedings of the 1st Int. Energy Transfer for Electric Vehicles Conference (E|TEV): Nuremberg, Germany, 2012.

6-9. C. Koebel, "PRIMOVE—Inductive Power Transfer for Public Transportation," Proceedings of the 1st Int. Energy Transfer for Electric Vehicles Conference (E|TEV): Nuremberg, Germany, 2012.

6-10. J. Y. Jang, J. Dae Ko, and S. Jeong, "Optimal design of the wireless charging electric vehicle," Electric Vehicle Conference (IEVC): Greenville, South Carolina, 2012.

6-11. W. Kuehn, "Digitale Fabrik—Fabriksimulation für Produktionsplaner," Carl Hanser Verlag: Munich, Germany, 2006.

6-12. n.a., "Abschlussbericht zum Vorhaben, Forschung und Entwicklung batteriebetriebener Schwerlastfahrzeuge (AGV), und deren Erprobung in einem Feldversuch im Container-Terminal, Altenwerder in Hamburg": Duesseldorf, Germany, 2011.

6-13. "Guetertransportleistung in Deutschland in Mrd. tkm 1991–2006," http://www.bmu.de/ fileadmin/bmu-import/files/pdfs/allgemein/application/pdf/verk_guetertransport.pdf, Accessed July 16, 2013.

6-14. F. Gerstenberg, M. Lehmann, and F. Zauner, "Elektromobilitaet bei schweren Nutzfahrzeugen," eb—Elektrische Bahnen—Elektrotechnik im Verkehrswesen 110 (8–9): pp. 452–460, 2012.

6-15. G. A. Covic, J. T. Boys, M. Budhia, and C. Y. Huang, "Electric Vehicles—Personal transportation for the future," 25th World Battery, Hybrid and Fuel Cell Symposium and Exhibition. EVS-25, WEVA, Shenzhen, China, 2012.

6-16. "Meeting environmental challenges: Electric motorways—a future possibility," Scania Newsroom, http://newsroom.scania.com/en-group/2012/11/09/meeting-environmental-challenges-electric-motorways-a-future-possibility/, Accessed August 14, 2013.

6-17. J. Franke, F. Risch, and A. Kuehl, "Produktionsprozesse für kontaktlose Ladeinfrastrukturen," ZWF—Zeitschrift für wirtschaftlichen Fabrikbetrieb 106 (5): pp. 361–365, 2011.

6-18. "The Future of Luxury Electric Mobility," Infiniti, LE Concept, http://www.infiniti.com/ us/models/future_models/ev.html, Accessed March 15, 2013.

6-19. "Advances Toward Production Readiness with Syncroness," Evatran, Plugless Power™, http://www.pluglesspower.com/plugless-power-advances-toward-production-readiness-with-syncroness,Accessed December 12, 2012.

6-20. "Plugless™ Level 2 EV Charging System (3.3kW)," Bosch Electric Vehicles Solutions, http://www.pluginnow.com/plugless, Accessed August 8, 2013.

# Chapter 7
# Wireless Charging of Consumer Electronics in the Automotive Industry

## 7.1 Wireless Charging of Consumer Electronics

As the number of mobile devices such as smartphones, tablets, and laptops increases, the convenience of our daily lives is enhanced. The improved capabilities of these devices, such as higher processing power, larger storage space, faster communication speed, and larger high-definition (HD) displays, help us enjoy always-connected, fascinating digital lives. However, due to the enhanced capabilities of these devices, the amount of power they consume is also increased. Thus, we are not concerned with the battery levels of these devices and carry secondary batteries or power adaptors.

To make our lives more convenient, several technical approaches have been developed: increasing the battery capacity, decreasing the required power consumption, and making power supplying or charging more convenient. The last approach, making power supplying or charging more convenient, is the topic discussed in this chapter: "wireless power transfer or wireless charging."

Wireless power transfer between two devices can be accomplished using several techniques, including inductive coupling, resonant coupling, and radiation. Besides the first two technologies, other wireless power transfer technologies are not adequate for use in the automotive industry because of their (relatively) long-distance, radiative characteristics. Thus, in this chapter, we briefly discuss the inductive and resonant coupling techniques for wireless power transfer and consider the criteria when adopting wireless power transfer technologies for consumer electronics in automotive.

## 7.1.1 Wireless Power Transfer Technologies in Hand

### *7.1.1.1 Inductive Charging Technology*

Inductive charging is a well-known technology that uses the inductive coupling between two devices. One, called a "charging station," sends electric power to the other device, which is usually a mobile device. The power transfer between the two devices actually occurs in the two inductively coupled coils, as shown in Figure 7.1. The charging station generates an alternating current in the "sending" coil to create the (electrodynamic) alternating magnetic field. This alternating magnetic field can induce the EMF at the "receiving" coil that is located in the near field area of the sending coil.

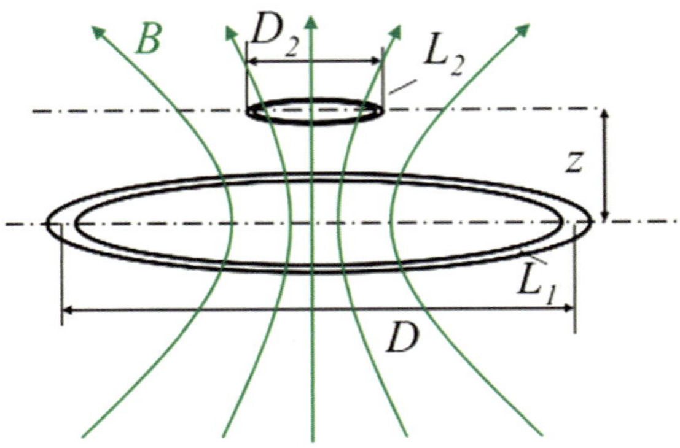


**Figure 7.1** Wireless power transfer between a pair of inductive coupled coils placed in close proximity [7-1] (courtesy of Eberhard Waffenschmidt).

The power transfer efficiency between these inductively coupled coils depends on the coupling coefficient $k$ between coils and the coils' quality factor $Q$.

As shown in Figure 7.2, when the quality factor $Q$ and relative size $D_2/D$ of these inductively coupled coils are given, as the distance $z$ between coils increases, the power transfer efficiency $\eta$ decreases rapidly due to the reduction of coupling coefficient $k$.

To increase the power transfer efficiency between two coils, most of the magnetic field generated at the sending coil should go through the cross section of the receiving coil. However, due to the nature of the magnetic field, as the distance between the two coils is increased, an increasing amount of the magnetic field does not go through the receiving coil. Thus, the effective power transfer distance of the inductive coupling method is limited within, at most, a few centimeters. However, the method has some pros: the effective EM field area is relatively small and the simple circuits give a stable/expected operation.

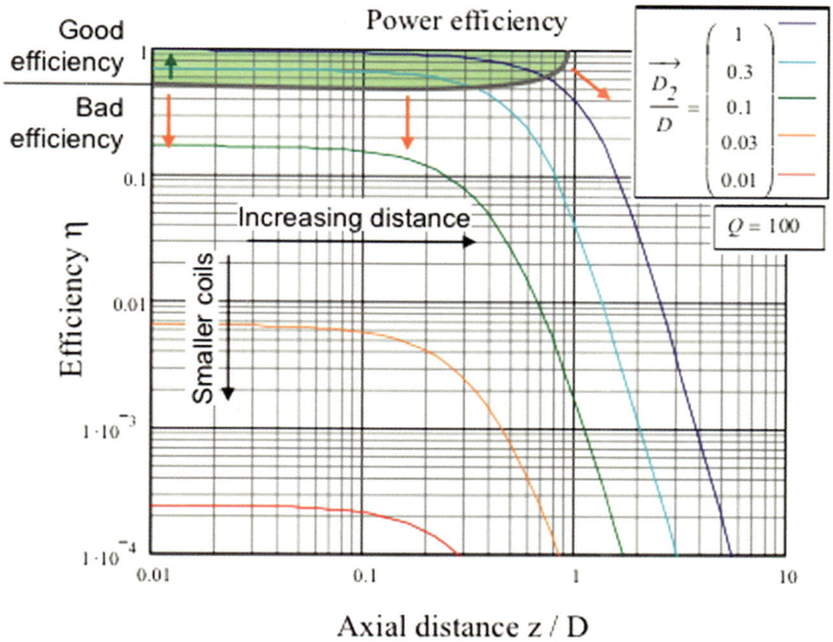


**Figure 7.2** The power transfer efficiency between inductively coupled coils depends on the relative size and distance of the coils [7-2] (courtesy of Eberhard Waffenschmidt).

### *7.1.1.2 Non-Radiative Resonance Charging*

First showcased by Prof. Marin Soljacic at MIT in 2007, non-radiative resonance charging is a special type of inductive coupling and is one of the technologies that can improve transmission distance. This method uses evanescent wave coupling generated between resonant coils within the near field; when the two resonant coils have an identical resonant frequency, we can observe resonance and energy exchange between them. Because this energy exchange is magnetic-field-based and has non-radiative characteristics, energy is transferred only when a coil has the identical resonant frequency, and energy that is not transferred is not radiated and is drawn back to the originating coil. Thus, this technology can diminish influences over other surrounding devices and near human bodies.

According to the coupled mode theory, the energy within the coupled coils results in the interaction shown in Equations 7.1 through 7.5. Between two coupled coils, the energy exchange pattern is as shown in Figure 7.3.

$$\frac{d}{dt}a_1(t) = -(i\omega_1 + \Gamma_1)a_1(t) + i\kappa_{11}a_1(t) + i\kappa_{12}a_2(t) \tag{7.1}$$

$$\frac{d}{dt}a_2(t) = -(i\omega_2 + \Gamma_2)a_2(t) + i\kappa_{22}a_2(t) + i\kappa_{21}a_1(t) \tag{7.2}$$

$$\text{in case}: \omega_1 = \omega_2 = \omega,\ \kappa_{11} = \kappa_{22} \approx 0,\ \kappa_{12} = \kappa_{21} = \kappa,\ \Gamma_1 = \Gamma_2 = \Gamma \tag{7.3}$$

$$\text{then}: a_1(t) = a_1(0)e^{-i\omega t}e^{-\Gamma t}\cos(\kappa t) \tag{7.4}$$

$$a_2(t) = ia_1(0)e^{-i\omega t}e^{-\Gamma t}\sin(\kappa t) \tag{7.5}$$

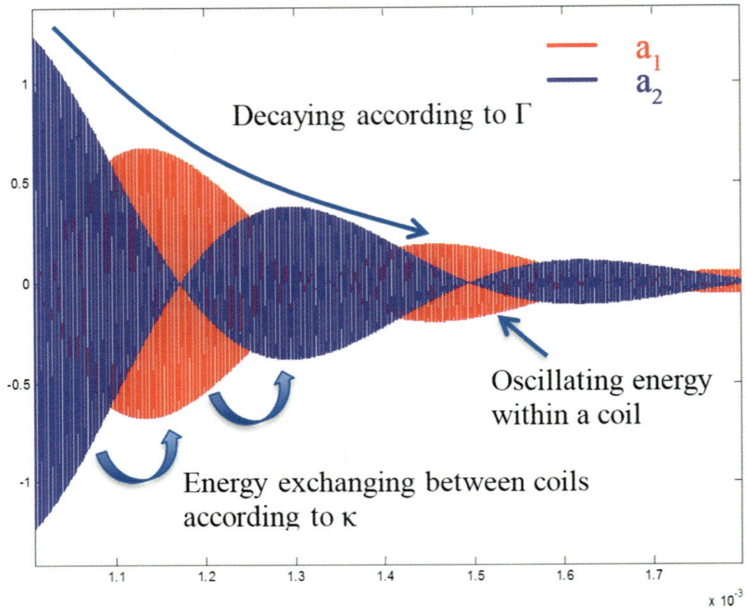


**Figure 7.3** An energy exchange pattern between evanescently coupled coils.

To use these methods, we need coils that have the same resonant frequency for energy transmission, and these resonant coils also require coils for supplying and collecting energy in them, as described in Figure 7.4. The reason we have to use separate supplying/collecting coils is that we need transmitting/receiving coils to have a high value of resonant $Q$ for evanescent coupling between them and secure safety and performance characteristics of the circuit.

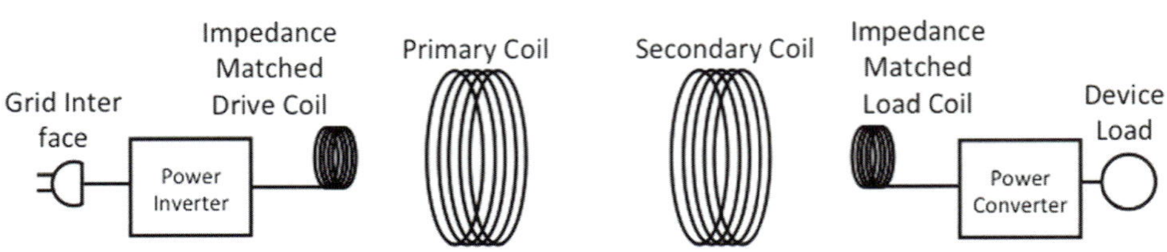


**Figure 7.4** A structure of a non-radiative resonance charging system. A pair of resonant coils is separated with drive and load coils for stable operation with high $Q$ resonant coils.

The scope of evanescent wave coupling is a critical factor that decides the range of performance of wireless power transfer technologies. This range depends on the resonant frequency and the size of the resonant coils, as described in Figure 7.5.

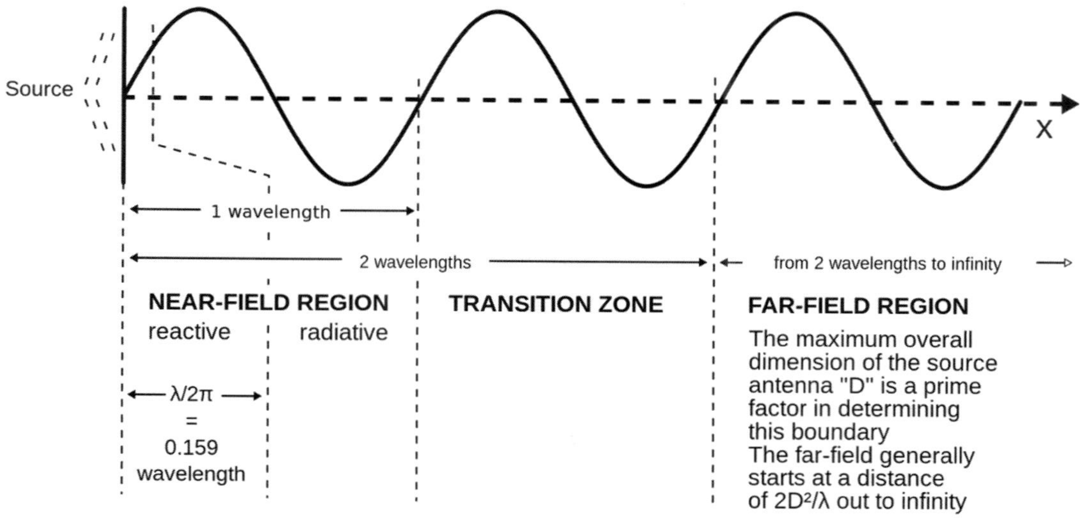


**Figure 7.5** A classification of electromagnetic field regions (figure from Wikipedia, original data from OSHA, USA) [7-3].

In companies such as Witricity, Intel, and Qualcomm, in-progress R&D aims for the commercialization of this non-radiative resonance wireless power transfer. The Qi standard of the Wireless Power Consortium (WPC) using this method was established in 2010. The consortium leads to the gradual increase of commercialization and application of the related technologies in the market.

In addition to its superior performance, efficiency, and price, non-radiative resonance charging is reported to have little effect on surrounding devices and humans. However, near the resonant coils for transmission, comparatively strong magnetic fields are generated, and users need to be careful with that. The simulated EM field is described in Figure 7.6.

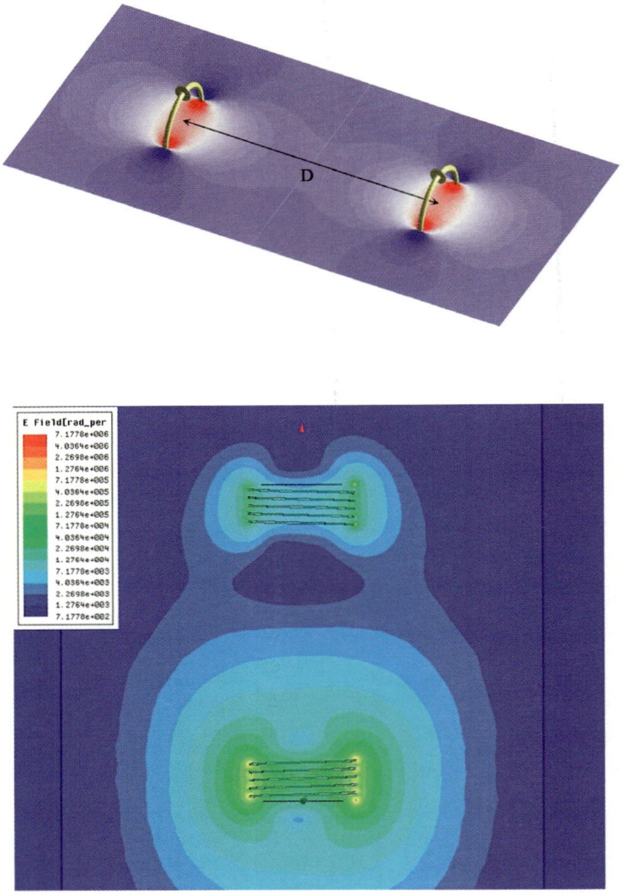


**Figure 7.6** The EM field intensities of resonance coils. The field intensity between coils is low, but the intensities at the close neighbor of coils are relatively high (top: © 2008 Elsevier. Reprinted with permission from [7-4], bottom: simulation results at KAIST).

## 7.2 Design Considerations

Before we adapt to devices in real life for several technologies of wireless power transfer that we have just discussed, we need to consider some important issues, given features of those devices. Not only everyday digital devices such as smartphones, smart pads, laptops, digital cameras, e-book readers, or Bluetooth hands-free products, but portable goods such as electric shavers, electric toothbrushes, and lanterns all use dc, rather than ac, as a power source, which means that they should have a (rechargeable) battery in them. This widespread use of dc is inevitable because there is no way to save or store power in ac systems. Therefore, adapting wireless power transfer to consumer electric devices means the existing wired rechargeable supplies in these devices have to be replaced by new power supplies using wireless power transmission.

Aside from which type of wireless power transmission we are using, we must take a few things into consideration for all devices: the first is wireless charging capacity, the second is wireless charging efficiency, the third considers the effects on surrounding devices, and the fourth is battery protection and battery life.

Wireless charging capacity must be able to supply power sufficient for recharging and running certain devices wirelessly. Therefore, wireless charging systems should be designed with the capacity of the existing wired power supply in each device taken into consideration. Most portable digital devices support charging via a USB port, and such devices can be charged with about a 5 V/2 A power supply. (A standard USB port is 5 V/500 mA, but some recent high-performance smartphones and smart pads run at 2 A for recharging.) Portable laptops normally require respectively more electrical power, so they mostly use 60–100 W power supplies. Some high-end laptops use even bigger power supplies, 150–200 W (19–24 V, 3–5 A).

When it comes to wireless charging efficiency, the "efficiency" here is in two ways: one is the wireless power transfer efficiency between the primary and secondary coils, and another is the system efficiency including running devices and restoring a dc power supply. That is, if the system efficiency of wireless power transfer is low, the subsequent power emission can increasingly affect the surrounding devices. Also, with low efficiency in driving and conversion, the considerable energy loss appears as heat. Plus, given the amount of the loss, we have to consider increasing driving electrical power for wireless charging devices according to the overall efficiency in wireless charging. Furthermore, we should consider how to cool the devices.

Effects on surrounding devices, in relation to certain devices and people besides wireless power transmitting devices, should be considered. Although basic regulations of EMI/EMF vary from nation to nation, we also have to consider electrical fields within a car. Recently, vehicles that were traditionally filled with mostly mechanical parts have increasingly been manufactured with diverse electrical parts, and those electrical devices and their performances should not be affected by the electrical field. The EMI/EMF issue is about not only direct effects on components but indirect effects delivered through the wiring inside the car body. Therefore, it is accordingly necessary to consider points such as distance, output, and frequency of wireless power transfer.

Battery protection and battery life issues are about effects on the performance of batteries that most mobile devices have in them. Short-range wireless charging technologies use changes in the near-field to transfer energy, and that change influences ions inside batteries. In an electromagnetic field, ions move according to their charges; this phenomenon is called eddy current. Movements of ions inside a battery cell not only can have effects on the aging of electrodes inside a battery cell but also can create heat inside it. So we need to minimize the aging and heat of a battery cell. For the minimization of such effects, the arrangement of a cell and an electromagnetic shielding structure must be designed in a way that minimizes the possibility of the battery cell being placed within the electromagnetic field.

## 7.3 Wireless Charging in Automotive Applications

Several points should be considered when we try to adopt wireless charging technologies in automotive applications due to the limited space of the vehicle and ownership of the power supply system and power pick-up system. In this section, we discuss each of these points.

### 7.3.1 Physical Safety

Due to the dynamic nature of vehicle movement and accidental situations, wireless power transmitters and receivers should be securely attached to the vehicle. The secure attachment of transmitters might be an easily solvable problem; however, for the receivers, the problem is not a simple one. The secure receiver attachment method might be one or some of retention clips, recessed cavities, or charging area covers.

### 7.3.2 Interferences

Besides the general EMI regulations, we should consider the additional interference effects with the in-vehicle modules, especially ECUs (ECM, TCM, BCM, and airbag controller), immobilizers (and remote keyless entry), TPMS, and radios. Most ECUs and their wiring harnesses are designed and installed to work properly in spite of severe external interference. However, their importance to vehicle operation and safety is critical. Thus, we should be concerned with how to minimize the interference to them. Additional EMI shields and sufficient distance between ECUs and wireless power transmitters can be considered when installing wireless power transmitters.

Interference to the immobilizers and TPMS might be more general than that to the CUs because these modules use an RF signal for their operations. TMPS generally adopts LF (125 kHz) and UHF (315, 433 or 868 MHz) wireless links for their operation. Even though the LF wireless links are generally located near the wheels (far from the typical location of the wireless power transmitter), we cannot disregard the interference possibility.

Similar consideration should be given for immobilizer systems. Most remote keyless entry systems use a UHF link for their communications. Thus, general EMI consideration (with an additional distance between RF receiver and wireless power transmitter) might be enough. However, an immobilizer system typically uses an LF (30–300 kHz) frequency for its communication and is located near the driver seat. Thus, more precautions should be taken. A wireless power transmitter should be located a minimum of 15cm away from the immobilizer system and immobilizer key. In addition, the wireless power transmitter should detect the foreign objects before and during the actual power transmission.

### 7.3.3 Vehicle Power Interface

Most passenger cars use 12 V dc as their internal power sources, but the voltage can vary between 9 V and 16 V due to the operation of vehicle electric power generators. Also, a 5

V dc USB charge connection may be available for portable smart devices. Thus, wireless power transmitter can be operated with 5–16 V dc power supplies. In addition, the damage to the in-vehicle power network should be minimized. Thus, excessive power consumption in wireless power transmitters should be limited by the use of a suitable fuse placed between the vehicle power network and wireless power transmitter.

### 7.3.4 Extremely High Temperature

Transmitters are expected to be installed in the passenger compartment. During the hot summer, the temperature inside a passenger car under direct sunlight can reach up to 78°C. Thus, the design and installation of transmitters should consider cooling and hot air ventilation and should verify the adequate operating temperature range limitations.

## References

7-1. Wireless Power Consortium, www.wirelesspowerconsortium.com, Accessed February 2015.

7-2. Eberhard Waffenschmidt et al., "Limitation of inductive power transfer for consumer applications," European Power Electronics (EPE) conference 2009, Barcelona, Spain, 8–10 Sept. 2009, paper #0607.

7-3. OSHA Occupational Safety & Health Administration, Electromagnetic Radiation: Field Memo, May 20, 1990, http://www.osha.gov/SLTC/radiofrequencyradiation/electromagnetic_fieldmemo/electromagnetic.html, Accessed February 2015.

7-4. A. Karalis, J. D. Joannopoulos, M. Solja i , "Efficient wireless non-radiative mid-range energy transfer," *Annals of Physics*, Vol. 323, Issue 1, pp. 34–48, April 27, 2007.

# Chapter 8
# Railway Application of WPT

In this chapter, the application of wireless power transfer to mass transit rail systems will be discussed. Mass transport systems include metros, subways, and other forms of light rail, high-speed rail, and magnetic levitation systems.

## 8.1 System Overview of Railway Application of WPT

Basically, a contactless power system for rail applications consists of a distributed primary winding installed along the track, which is supplied by ac current and at least one secondary pick-up coil that is used to supply the vehicle. A basic system overview is shown in Figure 8.1. The magnetic coupling between primary and secondary is usually increased by the use of ferromagnetic material in the secondary coil and, if necessary, in the primary coil. At rated powers of more than 100 kW, a working frequency in the range of 16 kHz–40 kHz is mostly used. The upper limit of the frequency is mainly given by the switching losses of the IGBTs in the inverter. On the other hand, the passive components and especially the coils get smaller with increasing frequencies at the same rated power. The basic overview of a contactless power system is shown in Figure 8.1. An inverter has to be fed mostly by a constant dc voltage. In some cases such as light rail applications, a dc-voltage bus is already available, so there is no need for a grid rectifier.

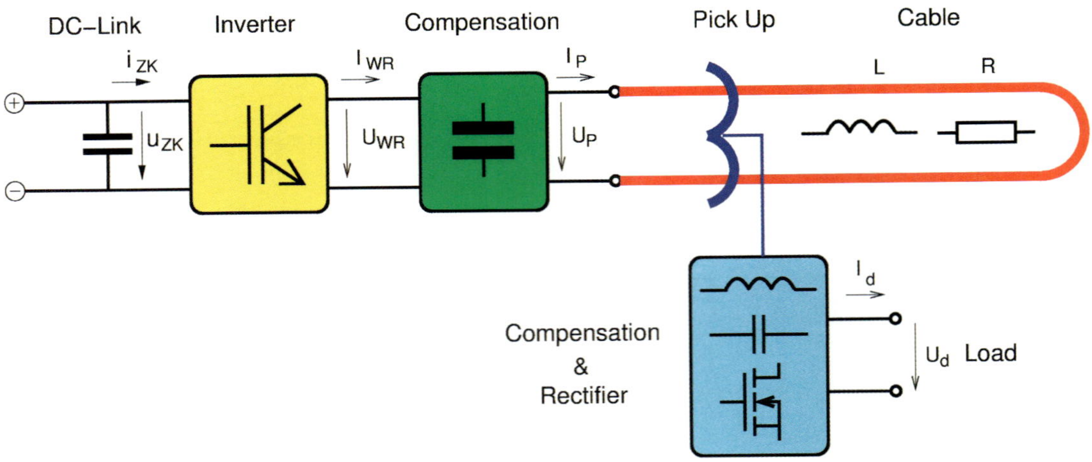


**Figure 8.1** System overview diagram.

The compensation network, which is termed a gyrator and is shown in Figure 8.2, converts the inverter ac-output-voltage into current, reduces output harmonics, and compensates the reactive power of the primary loop [8-1]. The resonant frequency of the compensated primary loop is set to the value of the inverter output frequency. So $\cos\varphi = 1$ is given for the fundamental oscillation of the inverter output. Under these working conditions, the switching losses are minimal.

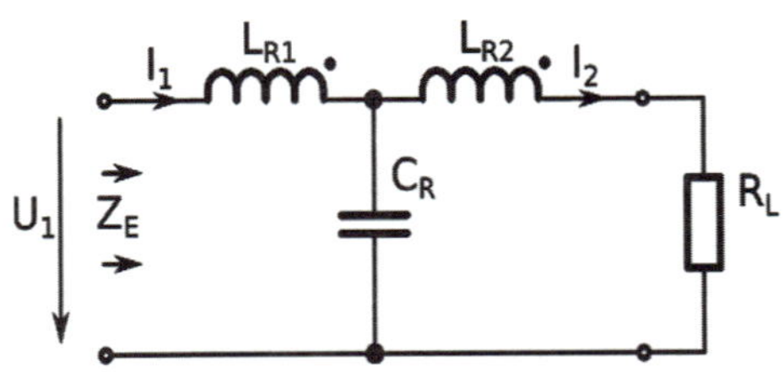


$$X = 2\pi f L_R = \frac{1}{2\pi f C_R}$$

$$I_2 = \frac{U_{1eff}}{jX}$$

**Figure 8.2** Gyrator circuit and its mathematical relations.

The resulting input currents at a square gyrator input voltage at both low and heavy load are shown in Figure 8.3. Due to the nature of resonant circuits, only the first harmonic is contributing to power transfer. The output current of a gyrator contains only the first harmonic sine shape and is theoretically free of higher frequencies. The latter occur in presence of parasitic impedances [8-2]. The transferred power increases in proportion to the primary current. In practice it should be considered that the efficiency is mainly defined by the primary losses and depending of the length of the primary. Usually a compromise between the value of the primary current and the pick-up design has to be made.

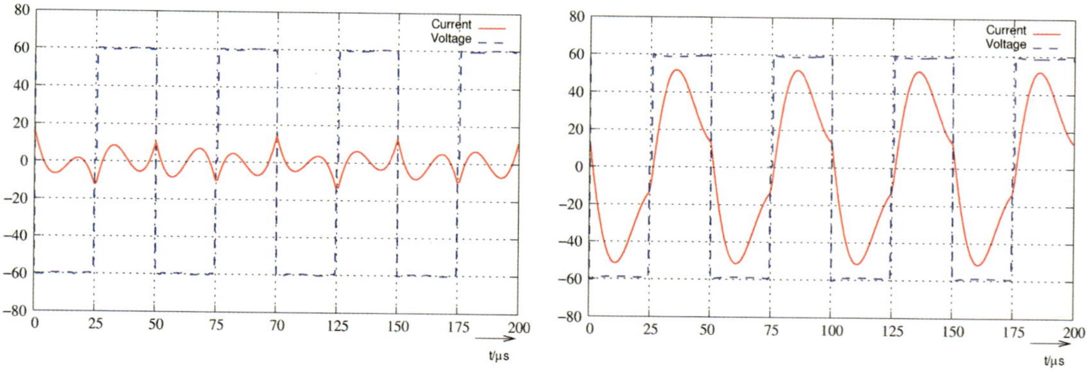


**Figure 8.3** Gyrator input voltage and current at no-load (left) and full-load (right) conditions.

Depending on the rated power, the primary current may be chosen in the range of 70–400 A. The primary current and generated output voltage should be in phase to set the power factor to 1. In conjunction with the compensation of the primary loop and the pick-up at the resonance frequency, the design of the primary inverter may be optimized to minimum size.

The use of frequencies in the range of 10–40 kHz requires special magnetic material for the pick-up iron core. Laminated iron sheets as used in conventional line transformers lead to unacceptably high iron losses. A good choice for the material of the pick-up iron core is a soft ferrite. Due to the limited size of soft-ferrite cores, some restrictions are given with regard to the geometric design of the pick-up coils. In the case of flat designs, a ferrite layer is required above the secondary coil windings. Its function is to increase the coupling and realize a magnetic shielding for the rest of the vehicle.

The distributed primary loop is installed along the track and is magnetically coupled to the secondary pick-ups. The number of secondary pick-ups is in accordance with the required power for each load and the number of individual loads. Each pick-up consists of an iron core with an assigned coil, a compensation network, and optionally a power converter for stabilization and voltage conversion.

In the case of a higher power need than one phase is able to deliver, because of voltage and current restrictions, there may be a higher number of phases installed in an independent or coupled magnetic design. Therefore additional primary wires have to be installed, including their compensation and the required power electronics. With such polyphased inductive power transfer systems, very high power density may be realized. Polyphased systems behave mostly as an equivalent single-phased system. Thus, further explanations will be restricted to single-phase approaches.

## 8.2 Track Segment Switching

The magnetic field produced by the primary coil along the track may be very high, especially for systems delivering high power. Because high oscillating magnetic fields may heat up metallic objects or harm living beings, the magnetization of the primary track should be switched off at the locations where it is not needed. Additionally, this saves energy and thus increases the overall efficiency of the system.

In the case of a short track or an unreachable location for living beings, the entire primary coil may be switched on and off as needed. If the primary coil is on the ground, as with most railway applications, a segmentation of the primary track should be done, where each segment may be energized separately. In this case, an approaching vehicle may be detected, and the related segments have to be switched on a short time before the pick-up arrives on the top of the primary coil. The possible approaches may be classified in two categories: center-fed systems and distributed systems.

The most important feature of center-fed systems is that a segmented track may be energized by a central inverter, so there is no need for mounting power electronics all along the rails. The inverters may be housed in distributing centers situated at long distances from each other. To reach redundancy, further inverters may be switched in parallel to the main inverter.

Center-fed systems may be current or voltage distributed. In the first case, only one gyrator is used to produce a constant current flowing in the feeder cables. Due to the resulting current source behavior, the segments are in series and get short-cut by a bidirectional switch, as shown in Figure 8.4 (top). A bidirectional switch may be mechanical like a contactor or on a semiconductor basis (e.g., two IGBTs in series). A segment is active when the respective switch is open, because the current may flow through the segment coil.

To minimize the voltage across the switches S1...S*n*, the resonant circuit including the inductance of the segment and its compensation capacitor should be well tuned to the system frequency. In this case, only the active voltage is present at short-cut time. The process of switching off the switches is more complicated because the voltage gets very high due to the current slope in the inductance of the segment. This voltage may damage the switch; therefore, a snubber circuit may be mandatory. Thus, the reliability of the system is low. However, this solution has the lowest costs and is adopted mostly for industrial applications in which only qualified persons are allowed to work with the system.

One of the major drawbacks of current-fed systems is that if the feeder loop is interrupted (e.g., if the cable is broken), the current source may produce an arc, which can't be detected easily. This issue may be solved by distributing the energy via a constant voltage source. This leads to a layout including a gyrator for each segment, with the switches located before the voltage-to-current transformation (Figure 8.4 bottom).

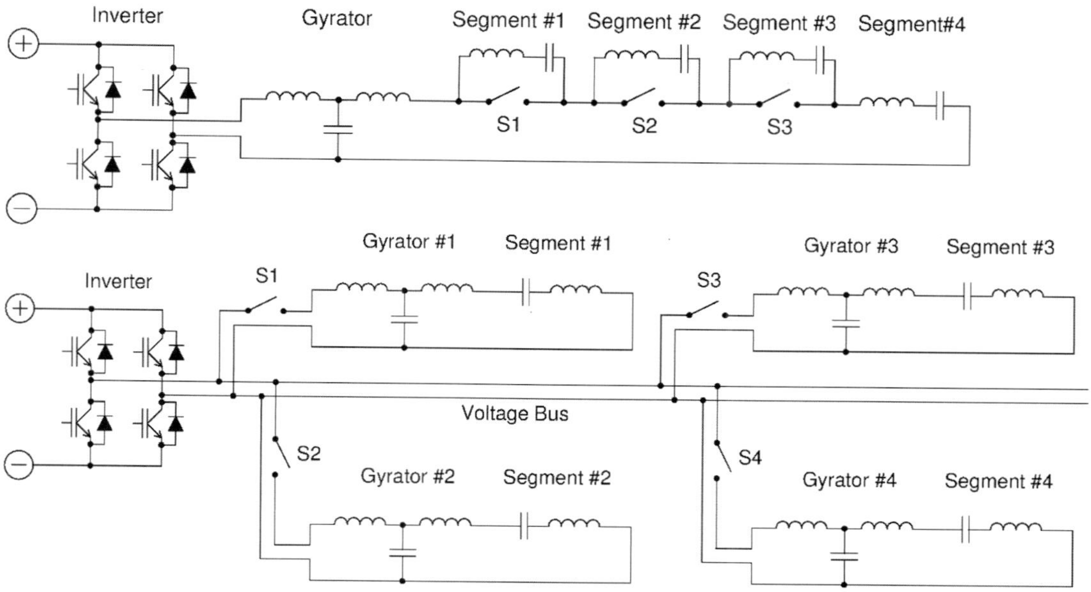


**Figure 8.4** Current bus center-fed system (top) and voltage bus center-fed system (bottom).

Depending on the rated power, the switched current may be lower than for current-fed systems. However, the voltage may also be very high at switching off, which would require snubbers to protect the switches as the segment under operation is turned off. Alternatively, the segment may be switched off at no-load condition. In this case, safer operation may be reached.

The voltage-fed system is based on a voltage bus cable along the track carrying a square-shaped voltage. Thus, all of the harmonic frequencies are present in a distributed way. This may lead to high earth currents and EMI/EMC issues, which may be solved by a coaxial or shielded feeder cable and an output filter at the terminals of the main inverter. Considering the square voltages on the primary coils of the voltage-fed system, the feeder cables should be buried far from the direct influential region of the electromagnetic field.

Center-fed systems include a switch per segment, which may be realized by semiconductors. Evidently a system with an inverter per segment may also be an advantageous solution. The switches are replaced by an inverter, which transfers the nominal or half of the nominal power, depending on the chosen topology. In this case there is no need for big stations, including all the electronics and electric interfaces. This simplifies the layout of tracks with little room for such installations.

On the other hand, higher redundancy is reached by distributing the power electronics units along the track, so that in the case of a failing inverter only one segment may be inactive, or the adjacent inverters take over.

Figure 8.5 (top) shows a system with completely separated four segments, denoted as Segment #1 to #4, each having an inverter and a gyrator, respectively. These may be considered as independent systems. However, the inverters should be synchronized to avoid cancellation and superposition of the primary field in case of vehicle overlapping at crossing segment borders. This approach is necessary and robust because of its independent channels and layout simplicity. This leads to higher reliability and adaptability. Its only disadvantage is the cost increase per segment, especially because each segment has to be designed to deliver the entire nominal power. A more economical configuration is presented in Figure 8.5 (bottom), connecting all four segments into a series connected segment system. This cost reduction can be achieved by arranging the segments in series and feeding each segment by its two adjacent inverters/gyrator combinations. Thus, each inverter may energize one half of the required power of two possible segments. The installed power capacity per inverter can be one half of the rated power.

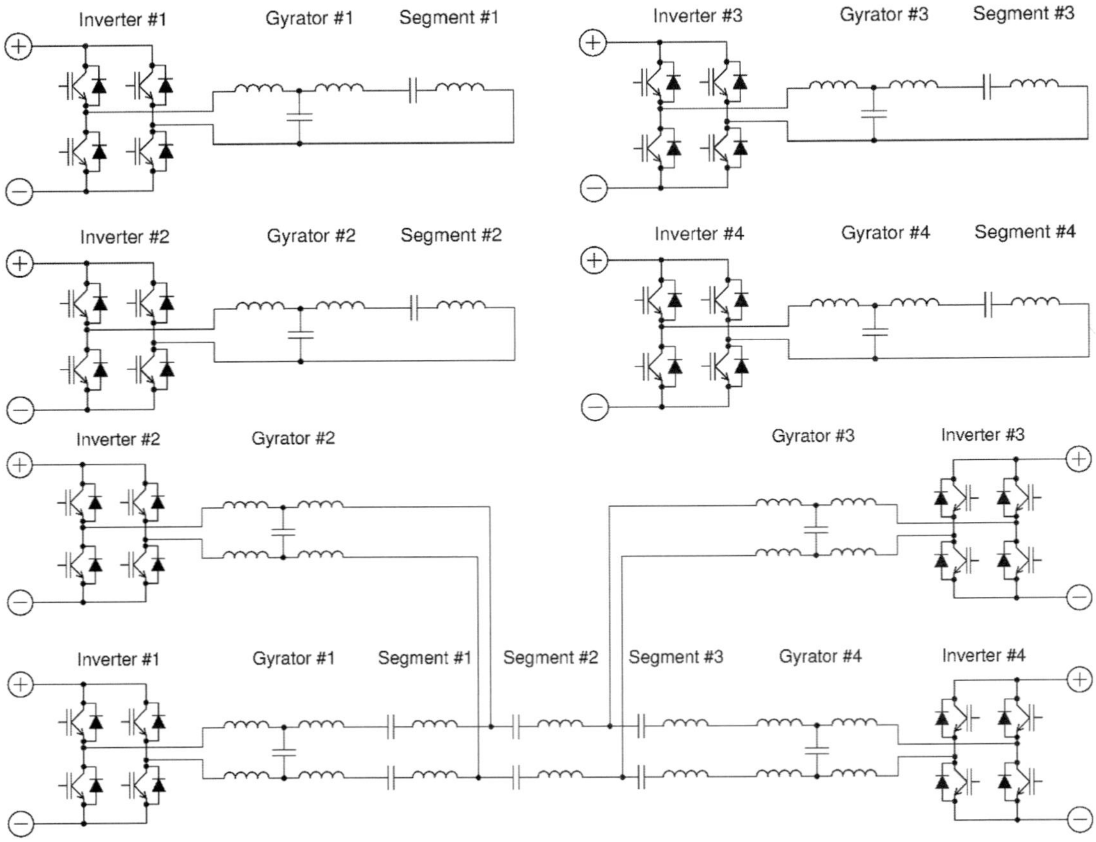


**Figure 8.5** Independent segment system of four segments (top) and series connected segments system (bottom).

The segment control unit has to synchronize each two adjacent inverters of the segment to be activated in a 180° phase shift, to obtain the highest available voltage at the terminals of the gyrators and thus the highest possible constant current in the segment. If the segments in the neighborhood of the active segment are not needed, the respective inverter at their ends should be short-cut by switching on their two low-side or high-side IGBTs. This short-cut is transformed by the belonging gyrator to a parallel resonant circuit, which has a high impedance at the working frequency and does not allow the current to flow out of the active segment.

To clarify the control sequence, an example may be considered for the configuration in Figure 8.5: assuming that segment #1 has to be energized, inverter #1 produces a square voltage with phase angle 0° at the terminals of gyrator #1, and inverter #2 produces a voltage with the same amplitude at phase angle 180°. If the series connection to segment #2 is ignored, the system is symmetrical toward the middle of segment #1. Theoretically, a short-cut may be done at the middle point of the primary coil without changing the behavior of the system. The current fed by gyrator #1 and gyrator #2 is constant at a phase angle of 90°. The series connection to segment #2 has to be ignored because the current fed by gyrator #2 should not flow through it. At the connection point, the circuit with the lowest impedance takes the highest current. To increase the impedance of segment #2, inverter #3 is shortened by turning on its low-side transistors. The inverter side of gyrator #3 builds a high-impedance parallel resonant circuit, and no current is able to go in that direction. The entire output power of inverter #1 and #2 energizes only segment #1.

When the vehicle is crossing the edge of two segments (e.g., #1 and #2), both covered segments must be synchronously energized so that the adjacent inverters #1 and #3 are activated with a 180° phase shift, and the middle inverter #2 at the intersection may be short-cut to build a high impedance and thus become inactive. This approach has a savings impact on the costs per segment, but it increases the complexity of the system. It is sensible to the tuning of the passive components, because the segments are dependent on each other. If one segment becomes out of tune, it may influence two segments, which have to be isolated by shorting the inverters of the respective neighbor segments. Compared to the independent solution, the series connection of the segments has lower redundancy and availability by half of the power electronics effort.

Finally, both solutions may be combined to get the best of both. An optimal design may be a compromise between the redundancy/availability and the costs of the system. A track with $N$ segments may be divided into $n$ independent groups of $m$ in series connected segments. Thus, $N = n \times m$.

## 8.3 Metro and Light Rail Applications

Metro rail lines such as subways generally consist of two or more rail car platoons that are individually connected or articulated with passenger access between cars. The metro line is energized by substations positioned every few kilometers along the tracks, including a power converter that reduces the distribution line voltage to 750V dc. This application scenario is shown in Figure 8.6.

**Figure 8.6** Light rail power substation, rail lines, and cars.

Figure 8.7 (a) shows dc power delivery via a third rail, or catenary, when above ground and in public spaces, that supplies multiple tracks. In this case, two tracks are energized for several miles from a point of common connection to the electrical distribution system. The station is equipped with circuit protection on both its utility distribution system side and on the rail, or load side. The step down transformer matches utility power to the operating voltage level of the metro line, generally 750 V dc, but in more congested areas, it can be 1100–1560 V dc [8-3]. During operating hours, the tracks remain energized, but individual sections can be isolated by isolation switches during maintenance or in emergency situations, and these isolation switches are shown as section switches in this figure.

A representative power profile for a light rail train is shown in Figure 8.7 (b), where Paux is the quiescent power needed to supply lighting, cabin air conditioning, and signaling. This chart is normalized, so the full acceleration and braking cycle is shown as 1pu (per unit) of time. During the interval 0 to TA1, the train accelerates to cruising speed at maximum power Pmx, reaching approximately 35–45 km/h, and then sustains this speed over the interval TA1 to TA2. Prior to braking, the accelerating (motoring) power is removed and the power delivery drops again to the auxiliary level before regenerative brakes are engaged at time TB1. During the braking interval TB1 to TB2, the rail car traction motors regenerate power back into the third rail (or catenary) or into brake resistors beneath the car. Dissipating braking energy to resistors is energy inefficient and, according to the authors in [8-4], can account for as much as 21% of the train's consumed energy (9000–23,000 kWh), or ~5000 kWh/day at the Korean Yangdong station and some 15,300 kWh/day at the Seolleung substation. This is substantial, so if wireless power delivery is proposed, it may need to be bidirectional.

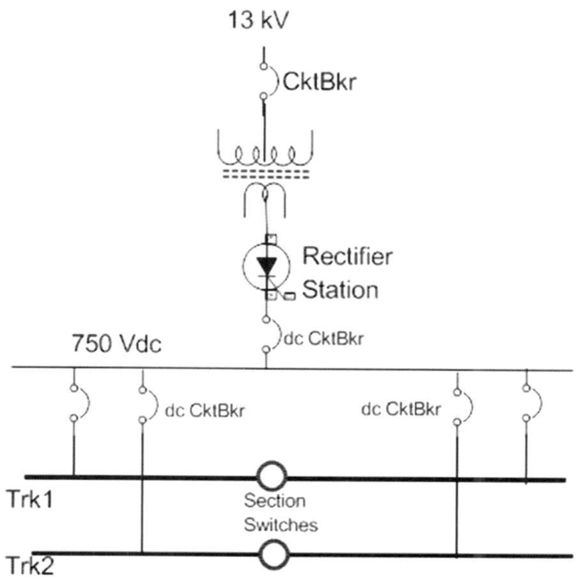


(a)

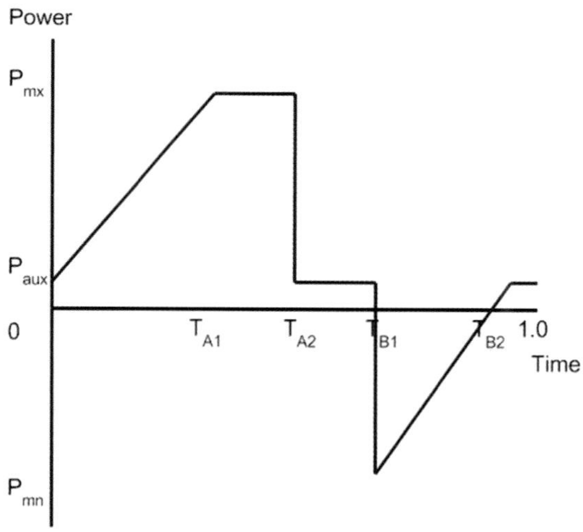


(b)

**Figure 8.7** Metro rail: (a) power delivery circuit diagram at a power substation, (b) power vs. time profile.

Wireless power delivery to light rail will, therefore, need to accommodate bidirectional power transfer and manage regeneration induced over voltages at the substations. At present, energy storage stations are placed at the power substations, at key locations along the route, or in some cases on the train cars themselves. It will be important for a

WPT installation to be compatible with such existing infrastructure, and, in particular, to have the capability of meeting the metro line peak power requirements of electric power and current, as shown in Figure 8.8 on a 750 V dc line for an eight-car train [8-5].

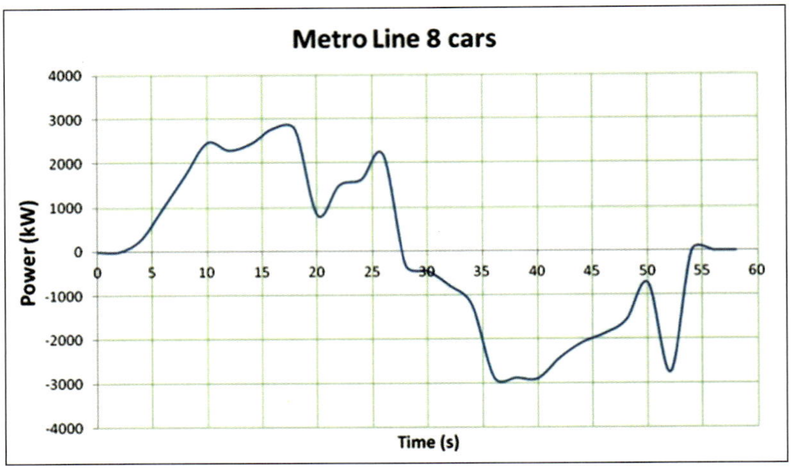


(a)

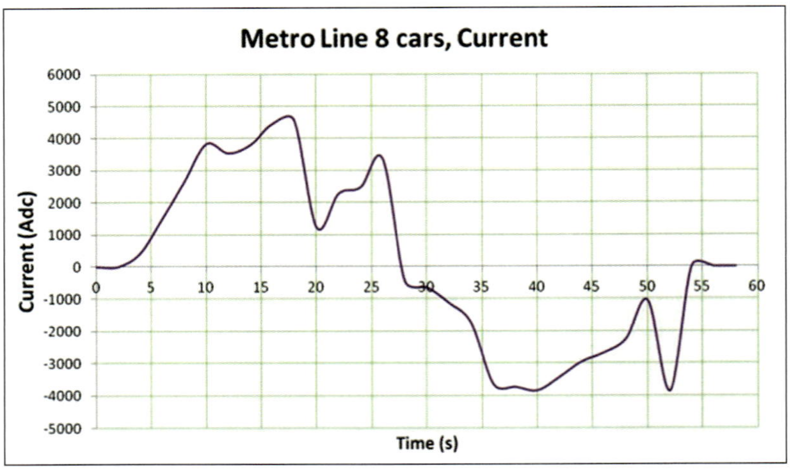


(b)

**Figure 8.8** Metro rail real power profile: (a) line power, (b) line current at a power substation.

The power level of an eight-car metro line is substantial, peaking at nearly 3 MW and the same during regeneration. Current levels at the substation are similarly high, on the order of 4000 A dc to 4500 A dc. At these power levels, it will benefit WPT power transfer to use higher system voltages. Figure 8.9 (a) illustrates the substation line voltage for the same 60-s interval of power and current shown in Figure 8.8. Note the fluctuation in voltage, sagging to 600 V at the 3 MW power peak at 17 s and surging to 800 V at the regeneration peak of 3 MW at 37 s.

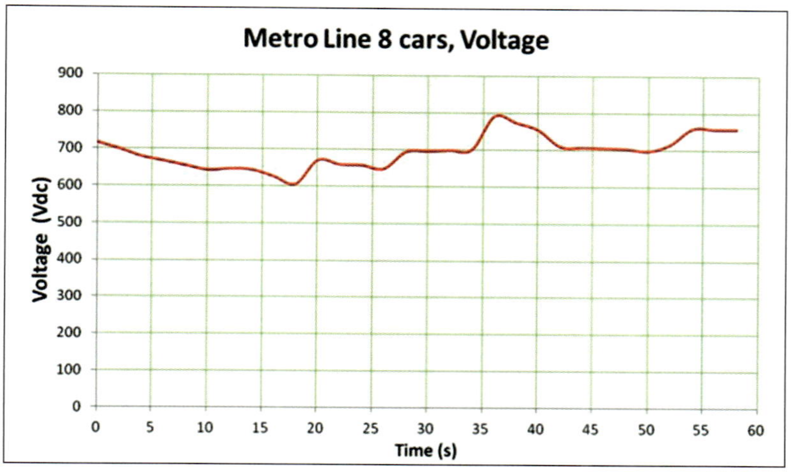


(a)

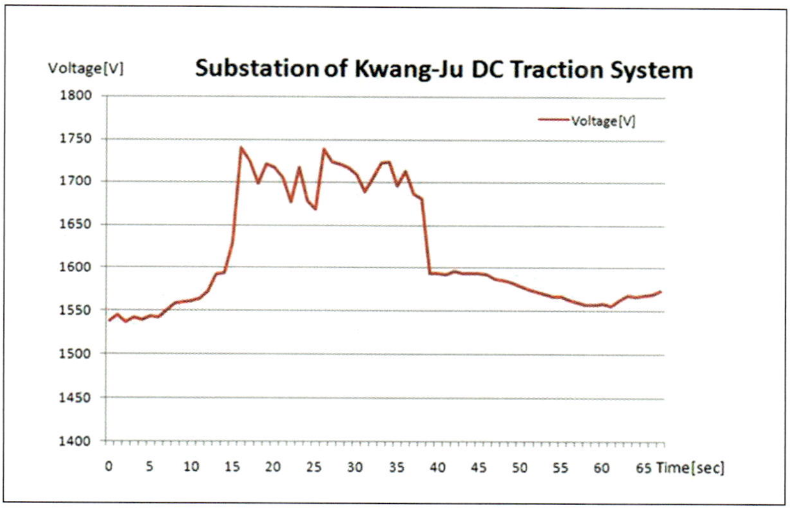


(b)

**Figure 8.9** Metro rail line voltage: (a) sag during acceleration and surge during regeneration on an example eight-car train, (b) measured voltage surge on a heavily loaded 1560 V dc metro line.

Similar, but much more severe, behavior is observed on very heavily used metro lines such as the Kwang-Ju line in Korea, which operates at 1560 V dc and during commuter car braking can peak as high as 1800 V and potentially damage the converter substation equipment, as shown in Figure 8.9 (b). This has led to the introduction of energy storage at the substations, as dedicated energy storage stations along the line or on the cars themselves. Figure 8.10 shows the line side energy storage equivalent of a metro line power substation.

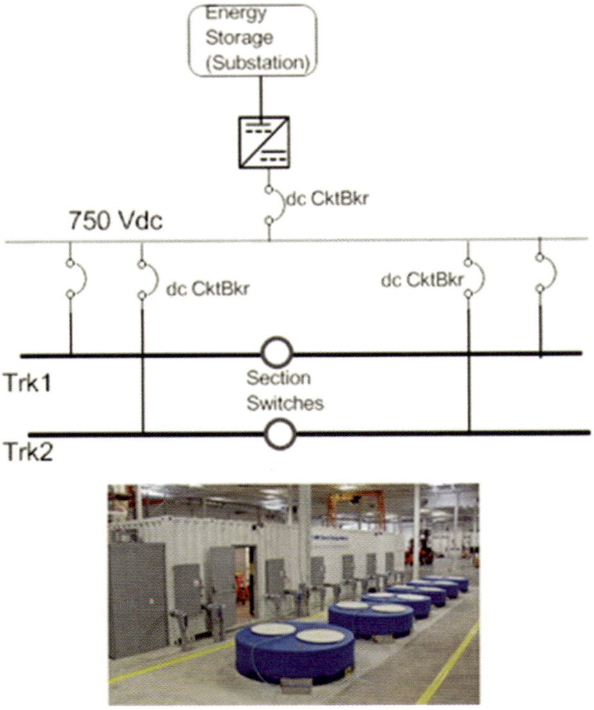


**Figure 8.10** Energy storage for metro rail line; Top: Architecture, Bottom: flywheel storage facility.

Light rail energy storage installations, such as that in Figure 8.10, must be rated for high peak power and charge and discharge durations of typically 25 s. Note the braking power duration for the eight-car metro line in Figure 8.8 (a) and (b) at the time of 25 s, and 3 MW on a 700 V dc line, and the same on a 1560 V line shown in Figure 8.9 (a) and (b), respectively, again at the time of 25 s. Metro line energy storage can be implemented in a variety of technologies such as advanced batteries, flywheel, and ultracapacitor, as well as hybrids of these. For example, a 200 MJ flywheel can deliver or absorb 8 MW for 25 s, which is sufficient to store the braking energy of a metro train without incurring line voltage surge and then can deliver the same to an accelerating train leaving the station without incurring a line voltage sag.

In these scenarios, the key point is that powering the metro train cars via wireless power transfer must be sufficiently rated and bidirectional. It is well known from dynamic WPT testing on light-duty EVs that high power peaks and valleys are typical of energized track sections comprised of individual coupling coils at a specified coil pitch. More will be said of this later in this section.

An alternative to trackside energy storage is to have energy storage components on the train cars themselves, as shown in Figure 8.11 (a) and (b), for a Bombardier train [8-6]. The pack consists of a series string of 2.7 V, 2700 F, carbon ultracapacitors for a pack capacitance of 4.5 F, a pack voltage of 1530 V, and total energy of 5.27 MJ (1.46 kWh). The packs are located on the train car roof, as shown in Figure 8.11 (a).

(a)

(b)

**Figure 8.11** Energy storage on metro rail car Mannheim, Germany: (a) Bombardier car, (b) Ultracapacitor pack.

More recent ultracapacitor applications in light rail in the United States include the TriMet district of Portland, OR, in which 7.3 mi of new line is being added to the existing 97-station Portland-Milwaukie line. This line will incorporate Siemens Sitras Static Energy Storage (SES) System and will be the first supercapacitor unit installed in the United States [8-7]. In another development, Southeastern Pennsylvania Transportation

Authority (SEPTA) has engaged ABB to introduce wayside energy storage systems (WESS) to their Market Frankford rapid transit line [8-7] using lithium-ion batteries. The first WESS installation was put in place in March 2013 and captures energy at five stations. Supercapacitors are used to maintain the voltage regulation of the third rail and to mitigate voltage spikes, and are being incorporated into the next phase of the WESS project that commenced in July 2013.

Based on this introduction of metros, subways, and other forms of light rail, it is clear that power levels involved will be in the range of megawatts (MW) rather than 10 s of kW for LD passenger vehicles. The ORNL has undertaken the development of dynamic wireless power transfer (DWPT) demonstration based on a small-battery electric vehicle. Figure 8.12 (a) illustrates the EV as it transitions on the electrified track consisting of a series of roadway embedded coils. The coils are placed at a 70% pitch in this design. Figure 8.12 (b) illustrates why, with nonpolarized coils, the power delivery will have peaks and valleys as the EV with a single centrally mounted secondary coil transitions across the roadway coils. A light pole affixed to the rear of the EV was used to visually indicate this variable power transfer. A similar variation in power occurs on the grid-side and is a reflection of the variation in a coupling coefficient as well as field reversals encountered by the moving secondary coil.

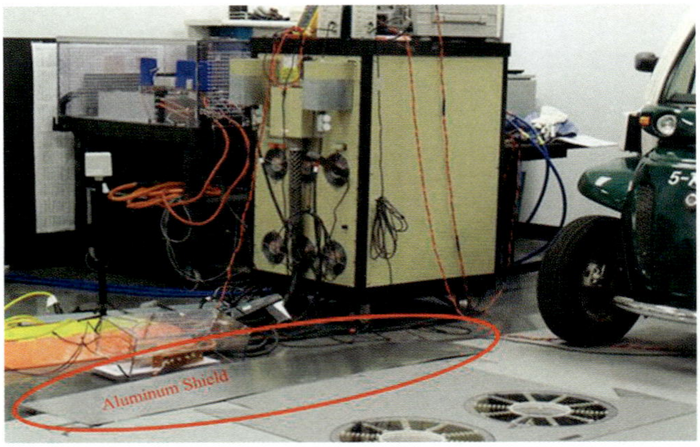


(a)

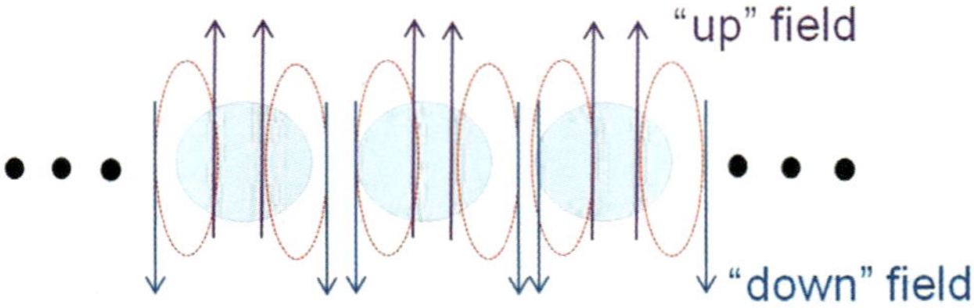


(b)

**Figure 8.12** DWPT demonstration: (a) Testing environment, (b) Nonpolarized field graphic.

An explanation of Figure 8.12 follows. In the experimental hardware, a single high-frequency power inverter operating at a 48 kHz central frequency had its power sequenced to the track coils in synchronization with the vehicle motion. During the course of testing, it was found that high-frequency magnetic fields caused by the current in the primary connection cables could potentially pose a field exposure hazard. To remedy this, an aluminum shield plate was placed over the connecting cables to minimize the magnetic field above these cables. The result was a substantial reduction in the fields, as discussed in detail in [8-8].

The graphic in Figure 8.12 (b) will help the reader appreciate the reason dynamically coupled power varies with location of the secondary coil. First, consider the case in which each of the track coils is wound in the same direction, for example, clockwise (CW). In this case, when energized, a coil produces an "up" field that varies sinusoidally with the primary coil impressed current. Therefore, in the inter-coil spaces, the field must return as a "down" field, again with the same sinusoidal variation. Now, consider a secondary coil of commensurate dimensions passing across these coils. When fully aligned with a coil, it will couple maximum power, and when aligned with the inter-coil space, this peak power will dip to some lower value that is dependent on coil shape and pitch. The condition of concern in this scenario is when the secondary coil straddles part of the up-field coil and part of the down-field inter-coil space. In this situation, the field is partially embraced by the secondary coil, which cancels, with the result being a minimum in power transfer. This condition occurs twice as the secondary coil moves across the inter-coil space.

Second, consider the case in which each of the track coil pairs is wound in alternating directions, such as CW followed by counter-clockwise (CCW), then CW, and so on. This is a case to be avoided because the inter-coil space magnetic field will be "flat." That is, the field from a CW wound coil will couple tightly with that of the CCW wound coil, resulting in a field that has a poor coupling coefficient once the secondary coils move out of full alignment with either the CW or CCW wound coils. Experimental testing at ORNL confirmed this phenomenon, and test data showed that power was at peak value when the secondary was fully aligned at either the CW or CCW coil but abruptly dropped as soon as the secondary coil moved out of alignment. This is a direct result of the "flat" field phenomena; the primary track coil coupling coefficient dropped abruptly once alignment was lost.

In Figure 8.13, the vehicle-mounted secondary coil encounters minimum power transfer during the inter-coil space (i.e., straddle points at positions #4 and #6) where up-field and down-field cancellation occurs. Note that this cancellation can be strong, such as at position point #6, where the coupled power is zero (i.e., complete cancellation). The remedy for power pulsations during DWPT charging of electric vehicles includes local energy storage as described for commercial light rail installations. The experimental demonstration of DWPT shown as Figure 8.12 (a) is depicted schematically in Figure 8.14, consisting of the GEM commercial EV with an 80 V lead acid battery pack. The

DWPT energized track consists of six coils connected in pairs and sequenced by the base unit controller. Radio communications between the vehicle and base WPT unit is needed for power flow regulation.

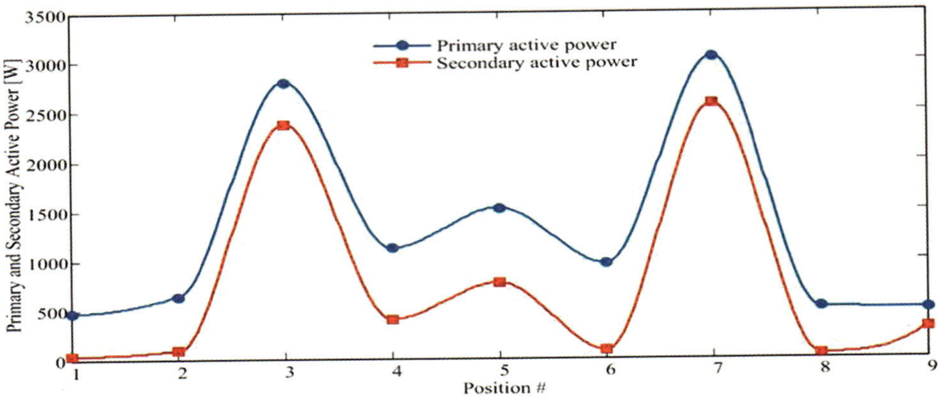


**Figure 8.13** DWPT power variations with position across a coil track. Coils fully aligned in positions 3 and 7; inter-coil space aligned in position 5.

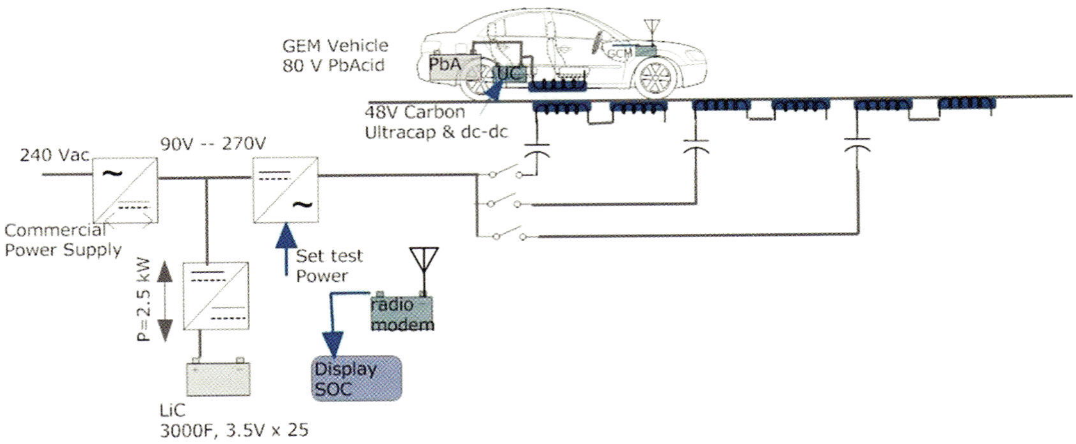


**Figure 8.14** DWPT showing vehicle- and grid-side energy storage (courtesy of ORNL).

In this demonstration, a commercial carbon ultracapacitor module (Maxwell Technologies 48 V, 167 F) is used to filter DWPT-delivered power prior to charging the vehicle onboard battery pack and from there to the vehicle traction drive electronics. Similarly, on the primary side, another energy storage pack consisting this time of higher-energy lithium-capacitor (LiC) technology is employed to smooth the grid power and thereby enhance power quality.

The system and technology shown in Figure 8.14 was exhibited at the International Transportation Electrification Conference, ITEC2013, held in Dearborn, MI from June

16–20, 2013. Figure 8.15 illustrates waveforms of the high-frequency ac current delivered by the power inverter shown in Figure 8.14 as the test vehicle drives over the WPT track. The charging power pulsations due to the passage of the vehicle secondary coil over the sequentially energized coils follows the pattern described in Figure 8.13, with short blanking intervals as coil pairs are switched. The difference in amplitude can be attributed to the vehicle lateral alignment with the track coils.

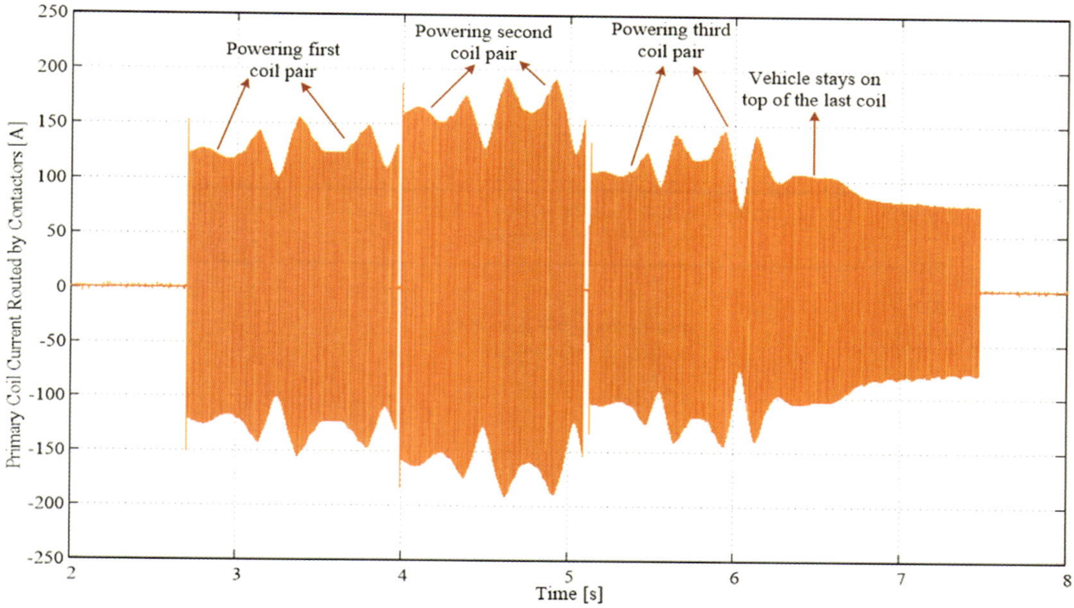


**Figure 8.15** DWPT track coil current during vehicle passage (courtesy of ORNL).

DWPT is viable for high-speed vehicles such as LR because coil energizing and de-energizing can be accomplished quickly via enabling and inhibiting the power inverter and then sequencing the coil contactors. Coil current during energizing follows the expected LCR (inductance-capacitance-resistance) circuit response to application of fixed ac voltage, as shown on the top of Figure 8.16. In this case, the 20 kHz excitation current contains a high magnetizing component necessary to magnetize the air in the space between coils and a component that transmits the real power. Because two primary coils are energized in series, the magnetizing component of current is higher than if single coils were switched. In a similar fashion, when a coil pair is de-energized, as shown on the bottom of Figure 8.16, the power inverter is inhibited, leaving only its antiparallel diodes to connect coil circulating current to the dc rail. Placing high-voltage dc across the coil results in quenching the coil's stored energy very rapidly, in the case of Figure 8.16, in approximately 0.2 s.

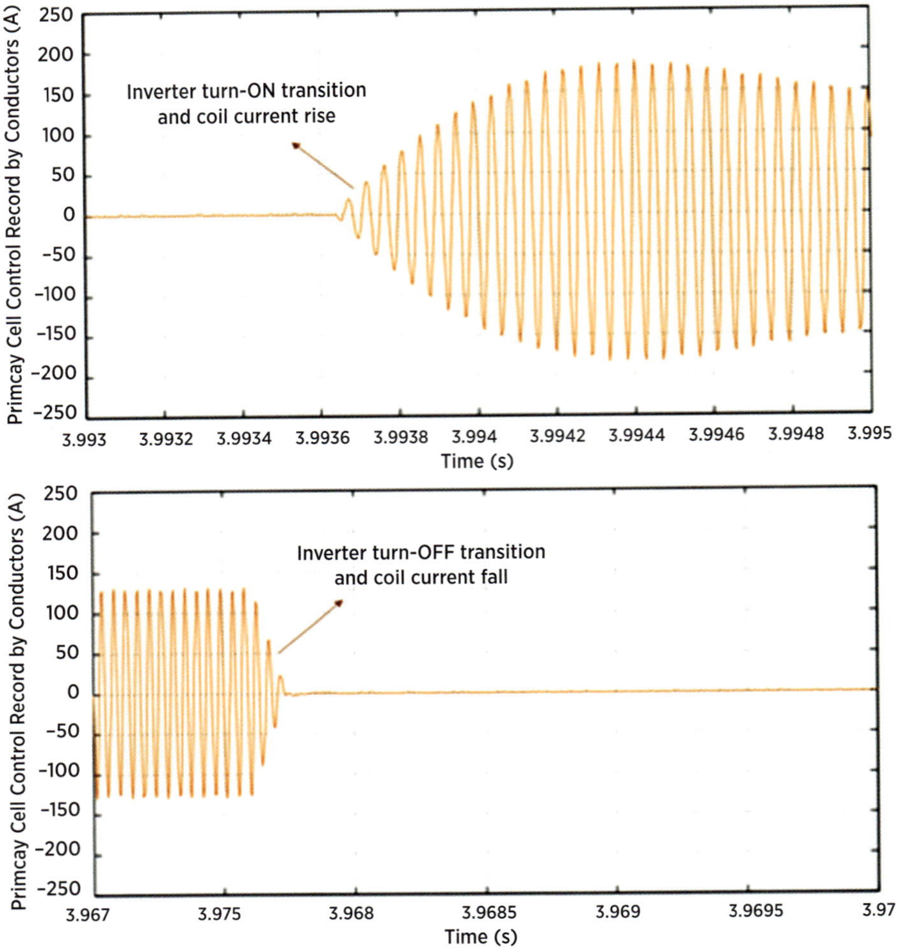


**Figure 8.16** Grid-side HF inverter current (courtesy of ORNL). Top: HF inverter current ring-up during sequencing, Bottom: HF inverter current quenching during sequencing.

The experimental results confirm that the main challenges of dynamic WPT are concerns over lateral alignment for lane keeping and optimum power transfer coupling, low-latency private and secure communications for V2I, highway construction and maintenance, utility power distribution to WPT roadside stations (RSU), time of use and revenue structure, and more. Not listed among these challenges are leakage fields, which for WPT are the magnetic, B, and electric, E, fringe fields associated with high-frequency magnetic resonance power transfer discussed in chapter 5, section 5.9.

## 8.4 TRANSRAPID Maglev Train

The choice of the long stator machine topology for the drive of the TRANSRAPID maglev has the consequence that there is no need to transfer the traction energy on the vehicle. However, 150 kW must be transferred to each vehicle section to energize the levitation and guidance actuators and the onboard supply system.

For driving speeds higher than 100 km/h, this required energy is delivered by a linear generator realized as a coil array mounted in the pole shoes of the suspension magnets. Due to the slot structure of the primary coil, the magnetic field is time-variant and induces a voltage in the generator coils. This voltage is rectified and fed into the onboard batteries. This assures the energy transfer to a vehicle driving fast, but the transferred power decreases at lower speed because of the relation of the induction to the field frequency.

The realized conventional solution up to 2004 to supply the train at low speed or at standstill, was the use of dc-sliding contacts and feeder rails. This causes wear and tear, high acoustic emission, and electric sparks, and it is sensitive to dirt and weather conditions.

A more convenient system was tested in 2006 on the Transrapid Test Facility Emsland (TVE), which consists of what is known as the Inductive Power Supply (IPS®) [8-9]. In contrast to sliding contacts, this system has no mechanical contact of the moving components, so that there is no dirt, weather dependency, or wear and tear. The power receivers are located all along the vehicle side, which allowed a distributed energy supply and a redundant operation thanks to independent systems situated on the two sides of the train. Further advantages are the galvanic separation and the relatively low short-cut power.

In this section, the system installed at TVE, in Figure 8.17, and its operation is explained.

**Figure 8.17** Test vehicle and track at the TVE (Transrapid Test Facility at Emsland).

The feeding system as shown in Figure 8.18 rectifies the three-phased grid voltage and converts the resulting 560 V dc voltage into a 20 kHz square voltage. The needed full-bridge inverter is realized by switching two three-phased inverters at 180° phase angle. Each of the three-phase legs of one inverter is connected in parallel to achieve the ability to switch higher currents [8-10].

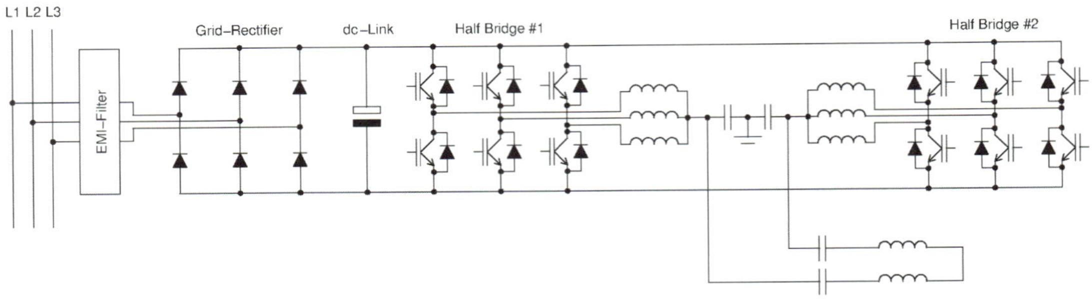


**Figure 8.18** Inverter circuit of an IPS-system.

The square voltage out of the inverter supplies a gyrator circuit to produce a constant 200 Arms current in the primary winding. This current is transported from the feeder cabinet to the track by a coaxial cable to reduce the feeder inductance and avoid magnetic fields in its environment (Figure 8.19, left).

Figure 8.19 (right) shows the prototype of an inverter, which feeds one side of the track with 250 kW [8-11].

**Figure 8.19** Prototype of the coaxial feeder cable (left) and the 250 kW primary inverter (right) (courtesy of IMAB TU-Braunschweig).

The constant primary current is led into two primary windings on the sides of the track. The inductance produced by these cables has to be compensated by series capacitors as depicted in Figure 8.20 [8-10]. These capacitors are also responsible for a moderate voltage against earth. Because the reactive voltage increases along the track, every 100 m a capacitor bank has to be installed. This way the voltage does not exceed the limit of 1450 Vrms at the rated current of 200 Arms.

**Figure 8.20** Primary track with compensation at TVE.

The pick-ups of the system are simple flat coils with a ferrite layer as backbone behind the copper Litz wires (Figure 8.21). This enables a low-cost, -volume, and -weight design to realize a high total power by cascading a high number of pick-ups.

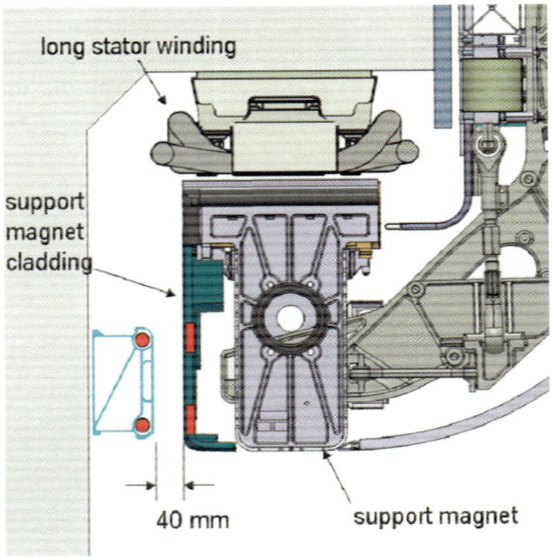


(a)

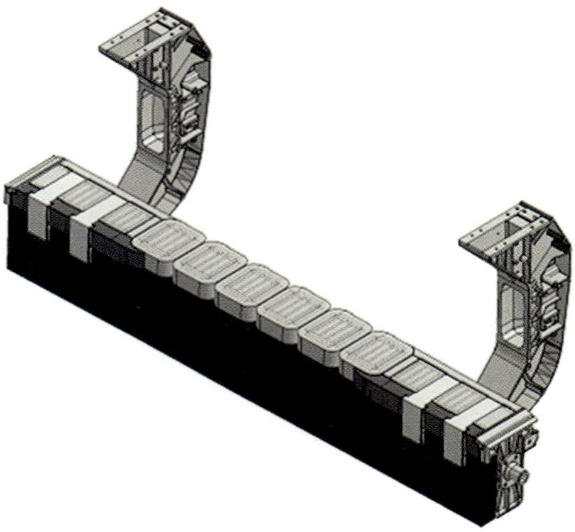

(b)

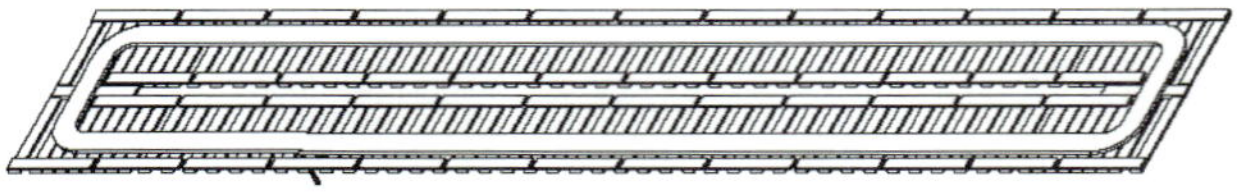

(c)

**Figure 8.21** Transrapid pick-up coil mechanical arrangement (a) side view, (b) assembled 3-D model, and (c) top view of the flat coils with a ferrite layer.

The secondary circuit on each TR-09 vehicle is built out of 15 pick-ups on each side fastened to the structure of the support magnets, as shown in Figure 8.21 (a) and (b) [8-12], [8-13]. These are classical series connection pick-ups with a rated power of 5 kW. The already existing 32 battery chargers in the TR-08 vehicle are reused to boost the pick-up output voltage (Figure 8.22).

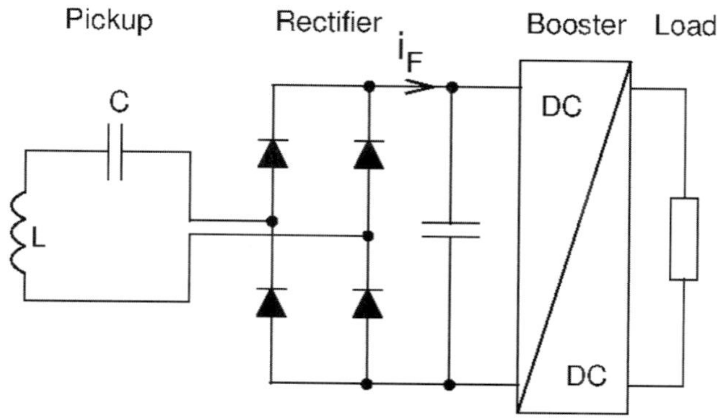


**Figure 8.22** Secondary circuit of the TR-09.

The test facility enabled a high number of measurements and research results regarding high power transfer. It showed that a total power of 500 kW may be delivered to a train. This value is up to now (2013) a still unreached benchmark.

After the dissolution of the joint-venture of Thyssen-Krupp and Siemens in 2005 and a severe accident in 2006, the TVE track has been closed in 2010 for Maglev tests, but it is placed at the disposal of further wireless power transfer tests.

## 8.5 Light Rail Application

The idea behind light rail wireless power supply application PRIMOVE is to create a catenary-free solution, without using any sliding contacts such as the third rail or switched rails [8-14], [8-15]. The latter are expensive and need high maintenance efforts.

Even though wireless power supply is more expensive at acquisition time, its low long-term costs may compensate for this to an acceptable lifetime cost.

The first prototype track was first running in Bautzen in 2008 with a rated power of 125 kW per pick-up and two pick-ups per vehicle. Its topology is a center-fed constant current layout, with five segments, as shown in Figure 8.23 [8-16].

**Figure 8.23** Test system with switched segments (courtesy of Bombardier Transportation GmbH).

Compared to conventional systems with a single-phase primary coil winding, the PRIMOVE system layout features a high transferred power density. This is reached by increasing the number of active phases to three. This technological development was already done with ac grid networks at 50 or 60 Hz. Assuming the currents in the three phases are symmetrical, a phase angle of 120° is built between each two phases. The sum of all the currents is equal to zero. Thus, there is no need for a cable for the opposite current direction as done for single-phase systems.

A further PRIMOVE system is realized in the city of Augsburg with a higher rated power of 160 kW per pick-up (Figure 8.24 and Figure 8.25). The system layout has changed to an independent segments distributed topology. This highly redundant system features a high number of inverters along the track. Each of them is able to deliver the entire 160 kW output power [8-17]. Due to the already available dc-supply given by light rail installation, each inverter is connected to the common 750 V dc bus. An input rectifier is not needed [8-18].

The inverters circuit is a typical B6-bridge as known in conventional inverters for three-phased electrical drives [8-18]. The gyrator circuit is built out of three equivalent single-phased gyrators. Only their absolute phase shift is different. Then the primary windings are just three cables connected to a star-point at the end of each segment. The resulting inductance of these cables is compensated by a series connection of capacitors located in the proximity of the primary power electronics, as shown in Figure 8.25.

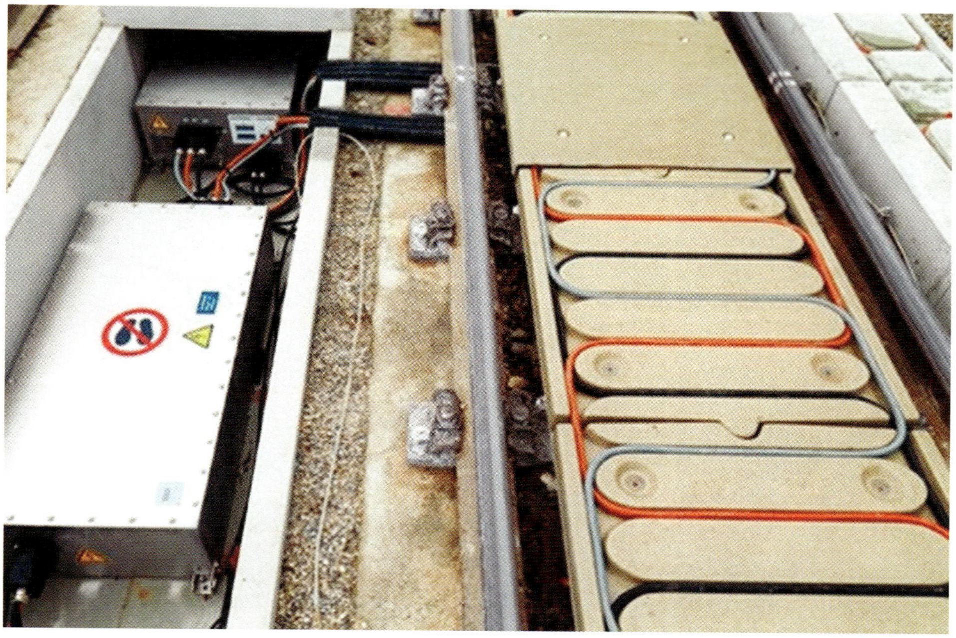

**Figure 8.24** Primary configuration with inverter box (left) and coil windings (right) (courtesy of Bombardier Transportation GmbH).

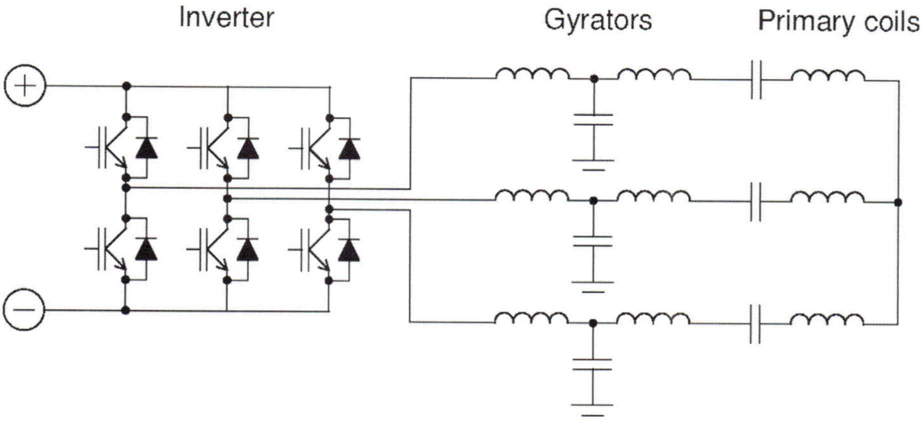


**Figure 8.25** Primary circuit (courtesy of Bombardier Transportation GmbH).

Each segment has to be energized at the time a vehicle is approaching. A system of vehicle detection is implemented to realize a reliable enable-signal for the primary field, as shown in Figure 8.26 [8-19].

**Figure 8.26** Vehicle components (courtesy of Bombardier Transportation GmbH).

In a similar way, the pick-up is also a star connection of three series resonant circuit partial pick-ups [8-18]. A delta-connection is also possible, but it may produce an unneeded loop current that doesn't contribute to the power transfer.

The coil array under the ferrite layer inside the pick-up features the same pole pitch as the primary windings. This leads to the best magnetic coupling. The induced voltage in the pick-up coils has to be rectified. A simplified rectifier circuit is shown in Figure 8.27.

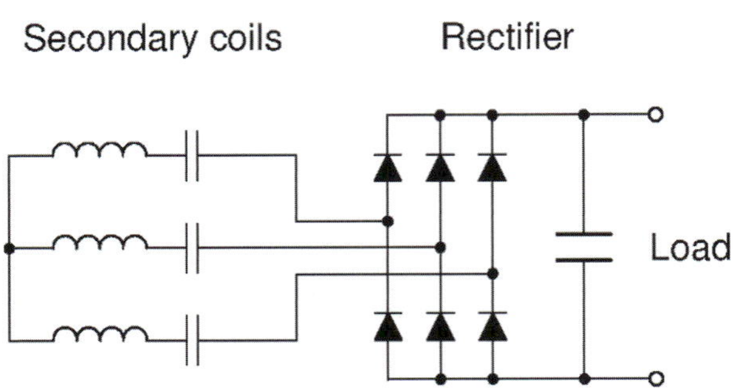


**Figure 8.27** Secondary circuit (courtesy of Bombardier Transportation GmbH).

Due to the dependency of the induced voltage on the magnetic coupling and thus the air gap between the primary and the secondary coil, the output voltage of the secondary rectifier is variable. As long as the voltage is in the input range of the drive inverters, the system is stable. However, some loads may need a constant voltage. Therefore, a dc-dc converter switched behind the rectifier may regulate the variable voltage. If the pick-up is feeding an energy storage system as a battery pack or an ultracapacitor, the required voltage stability is given by the onboard chargers or by a direct connection to the energy storage medium.

## 8.6 Wireless Low-Floor Train

In Korea, wireless power transmission was applied to the railway. Korea's wireless low-floor train refers to a system that operates using battery power alone without a catenary after charging the battery to its fullest power by bringing a pantograph into contact with a catenary at the top in a railroad station. Though it contributes to presenting beautiful cityscapes because there is no need for power lines during its operation in the downtown area, it poses the disadvantage of being equipped with a large-capacity battery. To overcome this, since 2011, the Korea Railroad Research Institute and KAIST have been conducting joint research on the methods to apply wireless power transmission, and they succeeded in demonstrating wireless power transmission at the test line of Osong in June 2013. In the test, a round trip to and from the station with a length of about 800 m can be made only by using the power supplied through the wireless power transmission during a stop and while on the move.

The first model of a large-capacity (60 kHz) wireless power transmission system applied to the wireless low-floor train was designed with a current collection voltage of 800 V, feeding two 15 m segments, spacing between road surface and current collection module of 30 mm, feeding capacity of 200 kW, and overall power transfer efficiency of 86%. The economic efficiency of this system was further enhanced through weight and cost reduction of the feeding and current collection devices. It was achieved by reducing the size of current collection modules by 1/3 and the number of power conversion devices (inverters) by about 1/9, compared with those of the 20 kHz-based wireless power transmission system technology of the OLEV, as described in chapter 4, while applying similar wireless power transfer principles. The technical specification comparison of both systems is summarized in Table 8.1. In particular, this system applied the resonant frequency of 60 kHz, instead of 20 kHz used by the existing German Transrapid and Canadian Bombardier Primove systems. The threefold increase in operating resonant frequency helped to achieve the lighter weight and reduced the size of feeding and current collection modules. It also led to increased power transfer efficiency, as well as secured a high-frequency operation of 60 kHz and larger power capacity technology.

Figure 8.28 shows a wireless low-floor train equipped with the 60 kHz large-capacity wireless power transmission technology. A conceptual diagram of the wireless power transmission applied to a wireless low-floor train is shown in Figure 8.29. The wireless power transmission system is composed of a power supply that can provide high grid power, an inverter that changes this power to a high frequency of 60 kHz, a feeding device that converts the high-frequency power into a magnetic field in the air, a pantograph that changes the magnetic field in the air to high-frequency power, and a rectifier and regulator, which make the high-frequency power collected from the pantograph into a dc power. As shown in the schematic of the onboard electrical power systems in Figure 8.30, the pantograph applied to wireless low-floor train changes the power collected from a pick-up device to the constant voltage required for a low-floor train operation through a power regulator, and charges the battery using the onboard dc-dc converter. The wireless power transfer system applied to wireless low-floor train is used not only for charging batteries during a stop, but also for motor propulsion and auxiliary loads with collected power while on the move. Figure 8.31 shows the resultant effect of improving urban landscape by removing the conventional catenary system when the wireless power transmission device is applied to a wireless low-floor train.

**Figure 8.28** Wireless low-floor train in which the wireless power transmission system is applied.

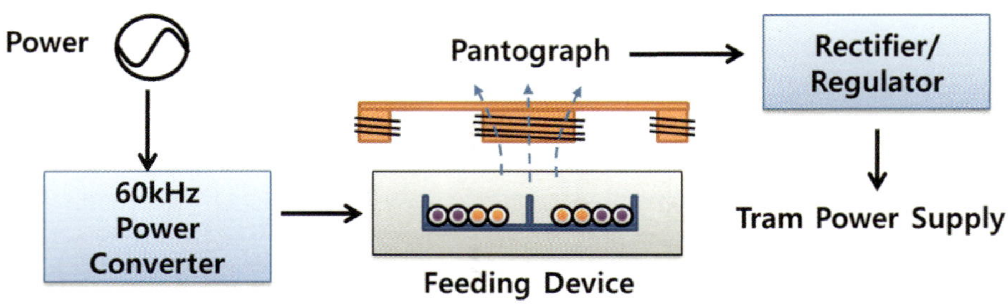


**Figure 8.29** Conceptual diagram of wireless power transmission technology.

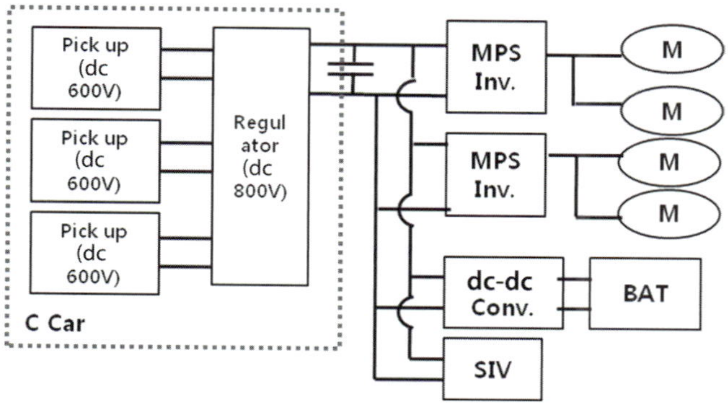


**Figure 8.30** Power system diagram of a wireless power transmission pantograph.

**Figure 8.31** Effect of improving urban landscape with the application of wireless power transmission.

The benefits that can be obtained by applying the wireless power transmission technology to railway vehicles are summarized as follows.

1. As the power required by the vehicle is supplied from the bottom of the train through a non-contact method, there is no need for maintenance and replacement of parts such as catenary equipment, and the maintenance costs can be reduced by 52% (maintenance cost per km) because maintenance work on the ground is possible.
2. Because catenary equipment is not required, the cross-sectional area of the tunnel is reduced by 20% when applied to the subway, and a 15% reduction in construction cost can be achieved (tunnel area of 56.0 $m^2$, 19.28 billion won → 45.05 $m^2$, 16.39 billion won) (Table 8.2).
3. As catenary and pantograph systems with less-efficient power transmission at high speed are not used, the development of the rail-type high-speed train is expected to be accelerated.
4. Due to the disappearance of the catenary on the ceiling of the train, the height can be increased up to 1.3 m in multi-story structural design of carriages and wagons, and accidents caused by feeder lines can be prevented, thereby securing the safety around the track.
5. In case a 180 kW class wireless power transmission system is applied to a wireless low-floor train operated using batteries, the installation of the wireless transmission facility can lead to more economical operation because the capacity of batteries mounted on the vehicles is reduced to about 1/4 in 10% of the entire route (a current wireless low-floor train battery capacity of 162 kWh → 41.65 kWh).
   - Battery capacity that can operate 500 m through charging for 37 s when wireless transmission facility is installed at 50 m: 1.65 kWh.
   - Battery capacity that can operate for 12 km in case of an emergency: 40 kWh.

**Table 8.1 Comparison Between Wireless Power Transmission Systems for an Online Electric Bus and for a Large Railway**

| | Wireless power transmission system for online electric bus (20 kHz) | Wireless power transmission system for large railway (60 kHz) |
|---|---|---|
| Collector unit module capacity | 20 kW | 60 kW |
| Size of current collector | 1.273 $m^3$ | 0.472 $m^3$ |
| Number of inverters | 80 | 9 |

based on 180 kW power transmission (electric capacity required for train operation); section in which an inverter is installed: 10 km

6. Urban railways, including a wireless low-floor train in which wireless power transmission technology is applied, are operated by sharing roads with buses and automobiles, and they are especially convenient because transfer on the road is possible. In addition, because there is no catenary, urban environments can become more pleasant, and complaints about damage to the urban landscape due to the underground urban railways can be resolved.

**Table 8.2 Effect of Reducing the Cross-Sectional Area of the Tunnel Due to Large-Capacity Wireless Transmission Technology: a 20% Reduction in Tunnel Area, a 15% Reduction in Construction Cost**

| Segment | Heavy rail transit (based on double track) | |
|---|---|---|
| Feeding method | Rigid bar catenary | Wireless power transfer |
| Cross-section of the tunnel | 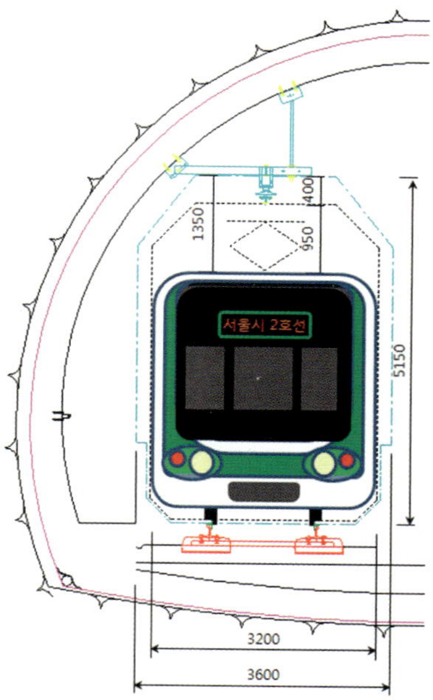 | 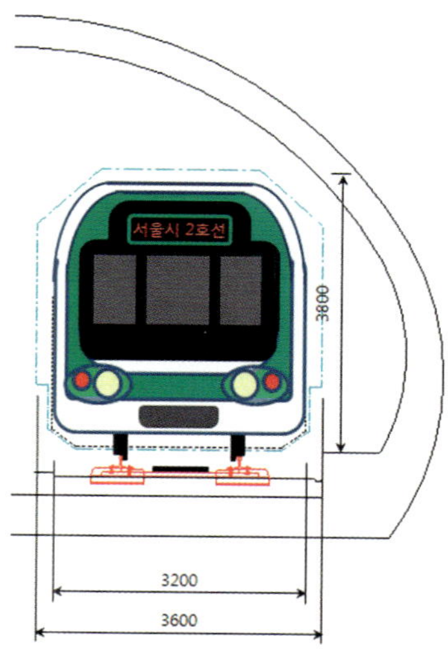 |
| Construction limits | W 3.6 m × H 5.15 m | W 3.6 m × H 3.8 m (1.35 m shorter) |
| Tunnel cross-sectional areas | 56.0 $m^2$ | 45.05 m2 (20% reduced) |
| Civil engineering costs | 19.28 billion won/km | 16.39 billion won/km<br>(2.89 billion won/km (15%) reduced) |

## References


8-1. J. Meins, R. Czainski, and F. Turki, "Phase control of resonant power supply inverters," Proc of EPE '05 Conference, Dresden, 2005.

8-2. F.Turki, Breitband-Datenübertragung für ein berührungsfreies ausgedehntes Energieübertragungssystem, ISBN-13: 978-3899596380; Der Andere Verlag, 2007.

8-3. M. Meinert, "Requirements on Energy Storage Systems for Railway Applications," Advanced Energy Storage Conference 2010, Session 3, Advanced Hybrid Solutions, San Diego, CA, 11–14 October 2010.

8-4. K. H. Cho, S. J. Jang, B. K. Lee, C. Y. Won, and G. D. Kim, "Development of DC Line Voltage Simulator for Control of Regenerative Energy Storage Device in DC Traction System," The 7th International Conference on Power Electronics, ICPE2007, October 22–26, 2007 / EXCO, Daegu, Korea.

8-5 J. B. Forsythe, "Light Rail/Rapid Transit: New Approaches for the Evaluation of Energy Savings, Part I—Life-Cycle Cost from Synthetic Routes/Operational Models," IEEE Transactions on Industry Applications, vol. IA-16, no. 5, pp. 655–665, September/October 1986.

8-6. O. F. Kashani, "IEEE IAS/PELS/IES Joint German Chapter meets in Mannheim," *IEEE Power Electronics Society Newsletter*, vol. 19, no. 4, Oct. 2007.

8-7. H. Glickenstein, "Worldwide Transit Upgraded: Art on Rails, New and Improved Systems, and Easier Payment," *IEEE Veh. Technol. Mag.*, vol. 8, no. 4, 13 December 2013.

8-8. O. C. Onar, M. Chinthavali, and J. M. Miller, "A Wireless Power Transfer System with Active Rectification and Recent Progress of ORNL WPT System Development," 2nd Conference on Electric Roads and Vehicles, CERV2013, Park City, UT, 4–5 Feb. 2013.

8-9. M. Bauer, P. Becker, and Q. Zeng, "Inductive Power Supply (IPS®) for the Transrapid," Thyssen Krupp Transrapid GmBH Munich, Germany Proc. Intern. Conf., Dresden, 13–15 October 2006.

8-10. G. Bühler, Systemverluste bei der berührungslosen induktiven Energieübertragung, ISBN: 978-3868531725, Dr. Hut, 2009.

8-11. J. Meins, G. Buehler, R. Czainski, F. Turki, "Contactless Inductive Power Supply," Maglev Conference 2006, pp. 527–535, Dresden, 2006.

8-12. A. Diekmann, W. Hahn, K. Kunze, and W. Hufenbach, "The support magnet cladding with integrated IPS® pick-up coil of Transrapid vehicles," Magnetically Levitated Systems and Linear Drives; Proc. Intern. Conf., Dresden, 13–15 October, 2006.

8-13. Q. Zheng, P. Becker, M. Bauer, and A. Diekmann, "Non-contact energy transfer for the on-board power supply of Transrapid vehicles," Thyssen Krupp Techforum, December 2005.

8-14. Bombardier Transportation GmbH: http://primove.bombardier.com/de/referenzen/augsburg, Downloaded August 12, 2013.

8-15. Alstom Group, http://www.alstom.com, Downloaded February 2015.

8-16. Patent WO002010000495: System and method for transferring electric energy to a vehicle, Bombardier Transportation GmbH, J. Meins and K. Vollenwyder.

8-17. MobiliCités, http://www.mobilicites.com/fr_actualites_les-tramways--les-bus-et-les-voitures-sans-fil-a-la-patte_0_77_1885.html, Downloaded August 12, 2013.

8-18. Patent US020120055751: Inductively receiving electric energy for a vehicle, Bombardier Transportation GmbH, J. Meins, C. Struve, and K. Vollenwyder.

8-19. Patent US020120318624: Detecting the presence of a vehicle, Bombardier Transportation GmbH, T. Dickson, K. Vollenwyder, and K. Berry.

# Chapter 9
# Long-Distance Power Transfer

## 9.1 Introduction

In the case of short distances, inductive or resonance coupling technologies with coils are effective. However, the distance when using coupling WPT is limited by coupling theory, even if we choose resonance coupling. To increase the WPT distance, we have to choose radio waves to carry energy. Although all WPT technologies are based on Maxwell's equations, their applications have minor differences. It is now possible to use higher-frequency radio waves (e.g., GHz microwaves) to focus electric power wirelessly at useful levels and increase beam efficiency to satisfy user needs. We use antennas instead of coupling coils to transmit and receive wireless power. It is also possible to transmit wireless power from multi-transmitting antennas to the multi-receiving antennas used in broadcasting and wireless communications because the antennas are not coupled electromagnetically. Systems that use very low power do not even require a battery and can be run on energy harvested from ambient RF and microwave radiation. The number of batteries required can be reduced when wireless power is widespread because they can be charged wirelessly, and hence, the shortage of batteries ceases to be a concern.

This scenario will soon become a reality. WPT via radio waves is, in fact, a valuable and convenient technology that can be used to charge batteries used in mobile phones, notebook PCs, EVs, LEDs, integrated circuits (ICs), and other equipment. In the near future, stable $CO_2$-free electric power from space will be possible with WPT technology. The concept of a space-based solar power satellite (SPS) is supported by WPT technologies via microwaves. In this section, the theory and application of microwave-based WPT are described.

## 9.2 History of Long-Distance Power Transfer

Historically, WPT began with radio waves. Maxwell's equations, formulated in 1862 and describing all phenomena of radio waves, are essentially the first theoretical basis for WPT. After Maxwell proposed his equations, Poynting described radio waves as an energy flow, and this is a well-known concept of the Poynting Vector. After the discoveries by Maxwell and Poynting, Nikola Tesla, well over one hundred years ago, dreamed that all electricity would someday be provided wirelessly. He conducted the first WPT experiment at the end of the 19th century [9-1], [9-2]. Unfortunately, he failed because of the diffusion of wireless power, which depends on the frequency of operation and the size of the transmitting antenna. He used an operating frequency of 150 kHz. After Tesla's failure, the history of radio wave development focused on wireless communication and remote sensing rather than WPT. However, the technologies for wireless communication and remote sensing helped lead to the development of new WPT techniques.

To increase the information contained in radio waves, it is necessary to use higher-frequency radio waves. The technology needed to produce a frequency higher than that used by Tesla, or microwave frequency, came into existence after World War II. Using a higher frequency with the same size antenna enables greater concentration of radio waves than is possible at lower frequencies. Therefore, the transmitted radio wave power can be increased with microwaves as a higher-frequency radio wave without the diffusion problem encountered by Tesla. In the 1960s, William C. Brown conducted new WPT experiments using high-efficiency microwave technologies [9-3], [9-4]. He used 2.45 GHz microwave tubes (such as magnetrons and klystrons) and succeeded in achieving WPT to a stationary helicopter in 1964 and to a free-flying helicopter in 1968. He was also the first person to develop a rectifying antenna, which he called a rectenna, for receiving and rectifying microwaves. The efficiency of the first rectenna, developed in 1963, was 50% at an output power of 4 W dc and 40% at an output power of 7 W dc [9-5]. Finally, he achieved an estimated 90% efficiency rectenna at 2.45 GHz. In 1975, he achieved a total dc-dc efficiency of up to 54% at 495 W dc by using a magnetron in the Raytheon Laboratory.

Simultaneously, Brown and Richard Dickinson's team succeeded with the largest microwave power transmission (MPT) demonstration in 1975 at the Venus Site of the Jet Propulsion Laboratory (JPL) Goldstone Facility [9-6]. The distance between the parabolic transmitting antenna with a diameter of 26 m and the rectenna array with an area of 3.4 $\times$ 7.2 m$^2$, was 1.6 km. The transmitted microwave power from the klystron source was 450 kW at a frequency of 2.388 GHz, and the achieved rectified dc power was 30 kW dc with 82.5% rectifying efficiency. This MPT experiment is the biggest ever conducted in the world.

While Brown succeeded with WPT field experiments, a large mismatch still existed between the demonstrated system size and cost and what would be practical. Therefore, unfortunately, commercial MPT systems did not become part of our daily life at the time. However, MPT technologies were developed for an SPS (Figure 9.1), as proposed by Peter Glaser in 1968 [9-7]. The SPS overcame some of the drawbacks of MPT, such as low overall system efficiency, which depends on microwave/dc conversion, and the large size of required antennas. The SPS was designed with an antenna of a few km, weight below 10,000 t, and 50% dc-microwave-dc total efficiency. The SPS supplied 1 GW stable electric power, approximately 10 times more than solar cells on the ground because it was in geostationary orbit (36,000 km above) in space and with the MPT. There is no nighttime in a geostationary orbit. In addition, unlike sunlight, microwaves are not absorbed by clouds and rain. Therefore, power generation is possible 24 h a day within an SPS. MPT was required for transmission of the power from the SPS, and consequently, MPT techniques were developed to meet the needs of the SPS [9-8].

**Figure 9.1** SPS Image.

After the 1980s, the active research region of MPT development moved to Japan, and some SPS research projects were conducted in Japan after the 1980s. In the 2000s, the SPS research project in Japan was still ongoing [9-9], [9-10]. On January 25, 2013, the Japanese government established a new basic plan for space policy that was revised from the old one established on June 2, 2009. In the new basic plan for space policy, four systems for utilization and three programs for R&D are described as a target for space development, including SPS technology. From 2009, when the old basic plan for space policy was established in Japan, The Institute for Unmanned Space Experiment Free Flyer, presently J-Spacesystems/Ministry of Economy, Trade, and Industry (USEF/METI) led a development program of a thin and high-efficiency phased array with Japan Aerospace Exploration Agency (JAXA). From 2012, JAXA started to design a small satellite for SPS application in parallel with the METI/USEF development.

For the SPS, Prof. Hiroshi Matsumoto of Kyoto University and his group conducted MPT rocket experiments in 1983, referred to as the Microwave Ionosphere Nonlinear Interaction experiments (MINIX). They also conducted several MPT field experiments on the basis of new microwave technologies for wireless communication and radar sensing [9-11]. An MPT experiment with a phased array was conducted in 1992 [9-12], [9-13], and this was the first trial to apply the phased array technique for this purpose. Kyoto University, Kobe University, and their teams flew a fuel-free airplane powered only with 2.411 GHz of microwave energy using a phased array with 96 GaAs semiconductor amplifiers and 288 antennas in three subarrays. The following year, the same phased array antennas were applied in the second MPT rocket experiment, called the International Space Year-Microwave Energy Transmission (ISY-METS). In FY2000, the SPS committee within the National Space Development Agency of Japan (NASDA), now called JAXA, conducted a phased array experiment with solar cells for SPS, which was referred to as the Solar Power Radio Integrated Transmitter (SPRITZ), a system developed mainly by Mitsubishi Heavy Industries, Ltd. The SPRITZ was designed on the SPS of the JAXA model.

The MPT can be applied to not only an SPS but also various commercial systems and others applications. In 1995, the same Kyoto University and Kobe University teams collaborated with a Japanese power company. They conducted a point-to-point MPT field experiment with a 2.45 GHz, 3 kW magnetron transmitter with a 3 m$\phi$ parabolic antenna and an approximately 3 m$\phi$ rectenna array. The distance between the transmitting parabolic antenna and the rectenna array was approximately 42 m. In 2009, Kyoto University proposed an emergency MPT system using an airship and conducted a field experiment (Figure 9.2) [9-14]. Two magnetrons on the airship at an altitude of over 50 m were used as 2.45 GHz transmitters. A mobile phone on the ground was charged with wireless power produced from the airship to simulate an emergency situation in which local power was unavailable.

(b)

(b)

**Figure 9.2** (a) Emergency MPT experiment in Japan. (b) Image of emergency MPT system.

## 9.3 Theory of Long-Distance Power Transfer

For long-distance power transfer, we do not use coupling coils; instead, we use an antenna for power transmission and reception. The beam efficiency between a transmitting antenna and a receiving rectenna is calculated by the Friis transmission equation, and its expansion is used for calculating beam efficiency in the near field. The Friis transmission equation is applied to the far field as follows in Equation 9.1.

$$P_r = \frac{\lambda^2 G_r G_t}{(4\pi D)^2} = \frac{A_r A_t}{(\lambda D)^2} P_t \tag{9.1}$$

where $P_r$, $P_t$, $G_r$, $G_t$, $A_r$, $A_t$, $\lambda$, and $D$ are the received power, transmitted power, antenna gain of the receiving antenna, antenna gain of the transmitting antenna, aperture area of the receiving antenna, aperture area of the transmitting antenna, wavelength, and distance between the transmitting antenna and receiving antenna, respectively.

The Friis transmission equation cannot, however, be used to calculate the received power in the near field because this equation assumes a plane wave in the far field, while spherical waves exist in the near field where WPT is used. Instead, the following $\tau$ parameter, given in Equation 9.2, is used to calculate the received power or beam efficiency $\eta$ [9-4], shown in Equation 9.3.

$$\tau = \frac{A_t A_r}{(\lambda D)^2} \tag{9.2}$$

$$\eta = \frac{P_r}{P_t} = 1 - e^{-\tau^2} \tag{9.3}$$

The beam efficiencies in the far and near fields calculated using the $\tau$ parameter are shown in Figure 9.3. The value of $\tau^2$ itself indicates the efficiency on the basis of the Friis equation. The beam efficiency can be increased to almost 100% when $\tau > 2$. To increase beam efficiency over long distances, the frequency should be increased. However, when the beam efficiency increases using higher frequency, circuit efficiency in a transmitter and a receiver and in a rectifying circuit generally decreases. The radio frequency-direct current (RF-DC) conversion efficiency of a developed rectenna is shown in Figure 9.4. Microwave is considered to be the best frequency range to achieve a high total efficiency that includes both beam efficiency and circuit efficiency. Hence, Brown selected microwaves and succeeded in achieving 54% dc-microwave-dc efficiency in the 1960s. It is expected that circuit efficiency will increase in the near future. If we can use higher frequencies with high-efficiency circuits, MPT will be more useful because the sizes of the antennas become smaller while maintaining the same total efficiency.

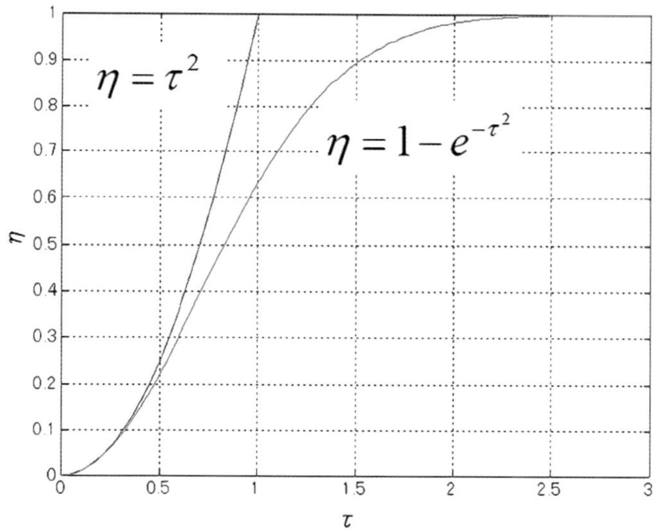


**Figure 9.3** Beam efficiency at the far field and the near field using the $\tau$ parameter.

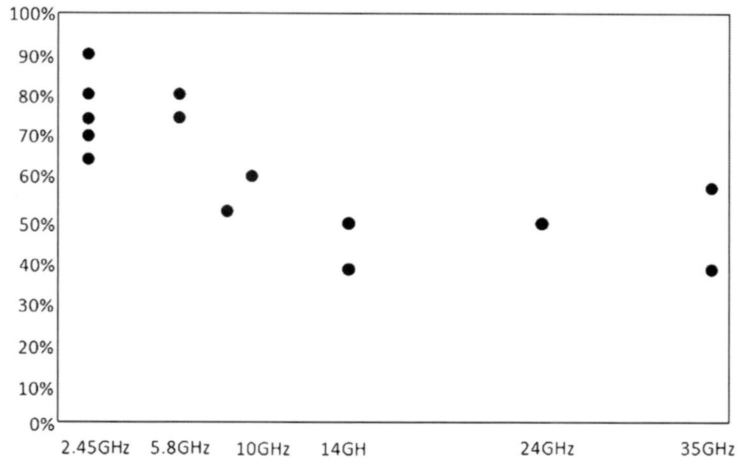


**Figure 9.4** Frequency dependence of RF-DC conversion efficiency of the developed rectenna.

This theory applies only to correctly positioned transmitting and receiving antennas. If the receiving antenna is not in front of the transmitting antenna (line of sight), the theory cannot be applied and beam efficiency decreases. The phased array, with which we can control the beam direction, should be used for MPT to a moving target to maintain high beam efficiency. After the MILAX experiment in Japan, some trials involving phased array MPT experiments (Figure 9.5) have been conducted [9-15]. We can apply a cooker-type magnetron for the phased array. A magnetron with stable frequency and controllable phase was developed originally by W. Brown [9-3] and replicated at Kyoto University. In Kyoto University, the magnetron phased array was developed for 2.45 GHz and 5.8 GHz operation [9-16].

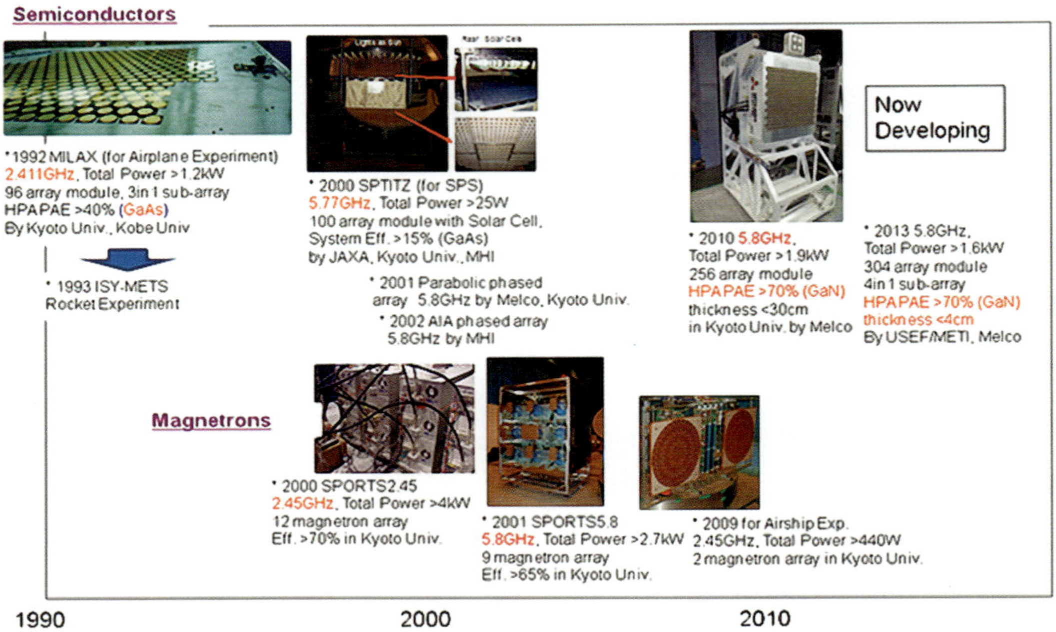


**Figure 9.5** Schematic diagram illustrating the history of phased array technology in Japan [9-15].

After the microwave power is collected, it must be converted to dc on the rectenna. The rectenna is composed of an antenna that receives microwave power, a low-pass filter that prevents higher harmonics from reaching the rectifier, and a rectifier with a diode(s) and an output filter to rectify the received microwave signal to dc. The ideal efficiency of the rectifying circuit is 100%. The rectenna typically consists of a single-shunt full-wave rectifying the circuit with one diode, a $\lambda/4$ distributed line, and a capacitor, which is typically used to reduce the loss in the diode. The rectenna can theoretically rectify microwave signals with 100% efficiency with only one diode [9-17]. A theoretical approach using a real diode was described by Yoo and Chang of Texas A&M University [9-18].

## 9.4 Recent Applications of Long-Distance Power Transfer

The following are suitable applications of the MPT.

- SPS (Super long distance power transfer)

  - Ten times larger power from solar cells in space (regardless of day/night, weather, etc.) including a total MPT efficiency of 50% (including beam and conversion, etc.)
  - Research topics: lighter weight, higher efficiency, size reduction, and low-cost phased arrays
- MPT to/from a moving target/transmitter (long-distance power transfer)
  - Impossible with wired power transmission and limited-time battery
  - High efficiency (> 90%) in long distance according to antenna theory
  - Research topics: higher efficiency phased arrays and utilization of higher frequency to reduce the size of antennas
- MPT for multi users with diffused radio waves or energy harvesting from broadcasting waves (long-distance power transfer)
  - Similar wireless system concept to information systems without major consideration of efficiency
  - Ubiquitous (everywhere and anytime) power source
  - Possibility of utilization of lower frequencies with higher efficiency because not using a beam system
  - Research topics: smaller- and larger-aperture antennas and high-efficiency low-frequency rectifiers
- MPT in a closed system or short-distance MPT systems
  - Highly efficient transmission with small antenna size over short distances
  - Free positioning because of the merits of MPT
  - Possibility of utilization of lower frequencies with higher efficiency
  - Research topic: higher power semiconductors

SPS and MPT to/from a moving target/transmitter were briefly described in the previous section. This section describes two scenarios: MPT for multi users with diffused radio waves or energy harvesting from broadcasting waves, and MPT in a closed system or short-distance MPT systems.

Beaming long-distance power transfer with high beam efficiency and diffused radio wave power transfer affect WPT via radio waves. Although the beam efficiency of the WPT with diffused radio waves is low and similar to that achieved by a broadcasting system, it is a convenient system. RFID is currently the only commercial application of WPT with diffused radio waves. This technology mainly uses the 900 MHz band.

The concept of a ubiquitous power source (UPS) was proposed in the 1990s and is based on the concept that microwaves are everywhere and are always available with WPT [9-19]. The concept of the UPS is shown in Figure 9.6. Microwave power at 2.45 GHz (industrial, scientific, and medical (ISM) band frequency in the 2.40–2.50 GHz range) is transmitted from the edges of a ceiling to charge mobile phones. It is easy to create a uniform microwave power density in a UPS room with antennas installed at the ceiling edges. Slot antennas are selected as the transmitting antennas because of their reduced cost. For the same reason, a frequency-stabilized and phase-controllable magnetron is used. However, the design of the UPS is limited by the safety issues associated with prolonged exposure to microwaves by human beings because the concept of the UPS is based on a "wireless power source all the time and everywhere." The safety level is under 1 $mW/cm^2$ for continuous exposure throughout the entire human body. Therefore, high-efficiency rectennas are necessary for operation at a microwave power density of 1 $mW/cm^2$ or less. Under these conditions, we have succeeded in charging mobile phones, as shown in Figure 9.7. Furthermore, a mobile phone can be used in the UPS room because of the difference between the 2.45 GHz microwave power and the 1.9 GHz communication system frequencies.

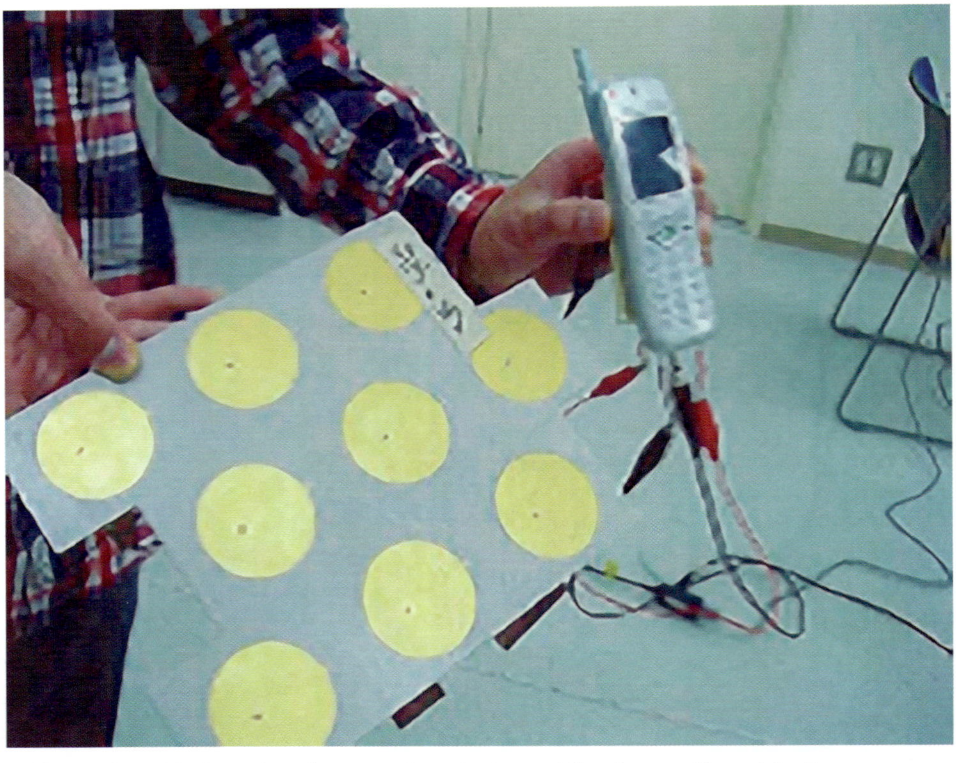

**Figure 9.6** Wireless charging experiment of a mobile phone with a ubiquitous power source [9-19].

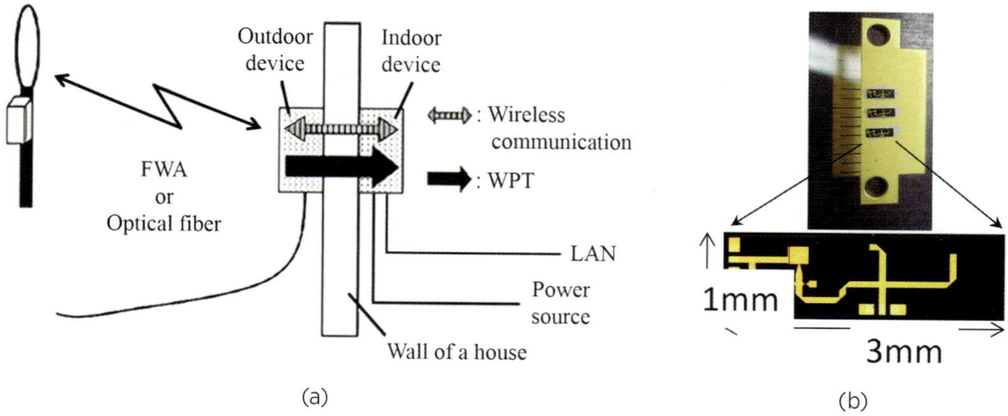


**Figure 9.7** (a) Concept of a microwave-driven FWA system (© 2011 IEEE. Reprinted, with permission, from [9-20]), (b) Developed 24 GHz MMIC rectenna (© 2013 IEEE. Reprinted, with permission, from [9-21]).

If users do not require a large amount of power, microwave power can drive their devices without a battery. One potential low-power device is a wireless sensor. The ZigBee/IEEE802.15.4 sensor was driven only with microwave power at Kyoto University in 2011 [9-22]. The ZigBee network is configured as a coordinator, router, and end device. The coordinator is just one device in a network and coordinates the network. The router has the function of relaying data from other routers and end devices as well as sensing information. The end device sends data only; otherwise, it sleeps. Therefore, power consumption of the end device is lower than the other devices. The router and end device send data for nearly 2 ms every 1.14 s. The coordinator decides to drop either of these from its network if it does not receive data for 15 s. The device out of the network must perform the necessary routines to rejoin the network. Hence, pulsed microwave power in the same frequency as ZigBee is suitable to increase microwave power with suppression of interference from ZigBee's wireless communication. Experiments involving a microwave-driven ZigBee sensor and its coexistence with a pulsed microwave power and wireless communication system on the same frequency of 2.45 GHz were carried out at Kyoto University and were successful, indicating that the systems could operate without interference in the same environment and at the same frequency.

The other proposed microwave-driven device is for fixed wireless access (FWA) as proposed by NTT Corp. and Kyoto University. Figure 9.7 (a) shows an image of the proposed system [9-20]. The outdoor device communicates through the Internet by the FWA or an optical fiber. The inside device and the outside device communicate wirelessly with each other. The inside device transmits power to the outside device by microwave. The outside device can operate with no battery. For the system, it is better that wireless information and power are carried by the same microwave carrier to reduce the size of the system. At first, 24 GHz was selected and an MMIC rectenna with a class-F load output filter was developed [9-21]. Dimensions of the MMIC rectenna at 24 GHz are 1 mm × 3 mm on GaAs (Figure 9.8 (b)), and the maximum RF–DC conversion efficiency is 47.9% for a 210 mW microwave input of 24 GHz with a 120 Ω load.

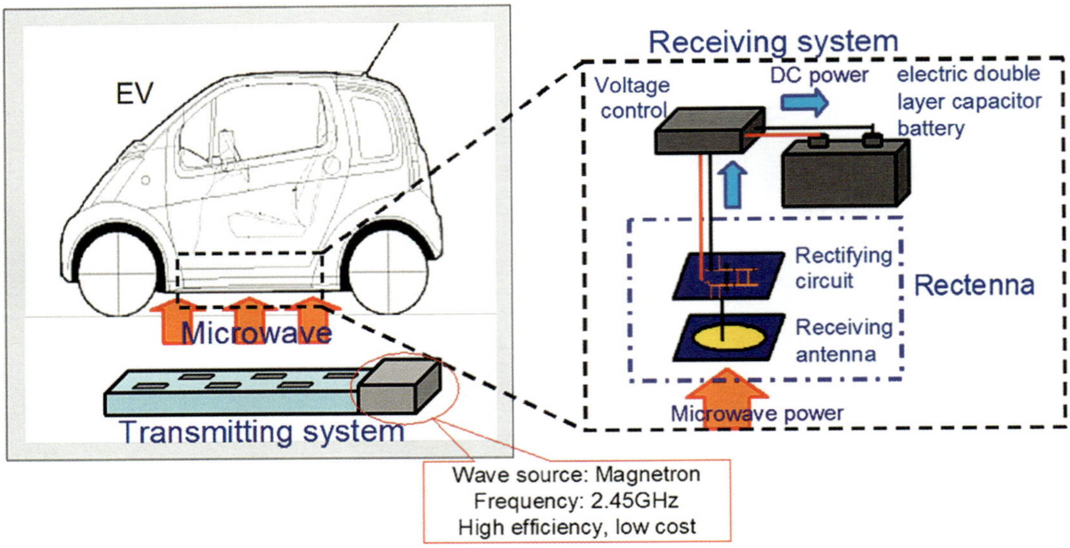


**Figure 9.8** Short-distance wireless microwave charging road system for EV [9-23].

In the near future, if the power requirements of mobile devices decrease sufficiently, wireless power can be received through broadcast service radio waves. Such technology would be considered a type of "energy harvesting" system. The definition of energy harvesting or "energy scavenging" is the process by which energy is derived, captured, and stored from external sources such as solar power, thermal energy, wind energy, salinity gradients, kinetic energy, and broadcast radio waves. One of the more well-known energy harvesters is the power microelectromechanical (MEMS) system, which can generate electric power from the fluctuation in a MEMS semiconductor. The energy harvester for broadcast radio waves would be a rectenna used for both MPT and WPT. A U.S. group used channel 48 (674–680 MHz) and harvested 60 $\mu$W (0.7 V) with a Yagi-Uda antenna [9-24], [9-25] at a distance from a broadcasting station of approximately 4.1 km.

MPT can also be applied in short-distance power transfer or power transfer in a closed system to reduce the size of antennas. In a short-distance power transfer system, wireless power can be increased to the kW range because there will be no humans or animals between the transmitting and the receiving antennas. One of the short-distance MPT applications is the wireless charging of an EV. Beginning in 2000, Kyoto University proposed and developed an MPT technique for an electric vehicle (Figure 9.9). Transmitting antennas were on a road, and rectennas were on the bottom of the EV. The distance between the transmitting and the receiving antennas was approximately 12.5 cm, 1 $\lambda$ of 2.45 GHz. The battery on the electric vehicle can be charged using only microwave transmission with a theoretical beam efficiency of 83.7% and an experimental beam efficiency of 76.0% [9-26]. This efficiency is high enough to economically transmit wireless power with microwaves.

Figure 9.9 is not the only MPT EV charge system. A short-distance MPT system also has merits and demerits. Merits include high beam efficiency with small antennas, and demerits include the occurrence of unexpected coupling between antennas [9-26]. To eliminate the demerits of unexpected coupling, we can expand the distance between a microwave transmitter and a receiver for wireless EV charging [9-27]. A phased array is suitable for a long-distance system to increase beam efficiency, to reduce side lobes, and to follow a moving car.

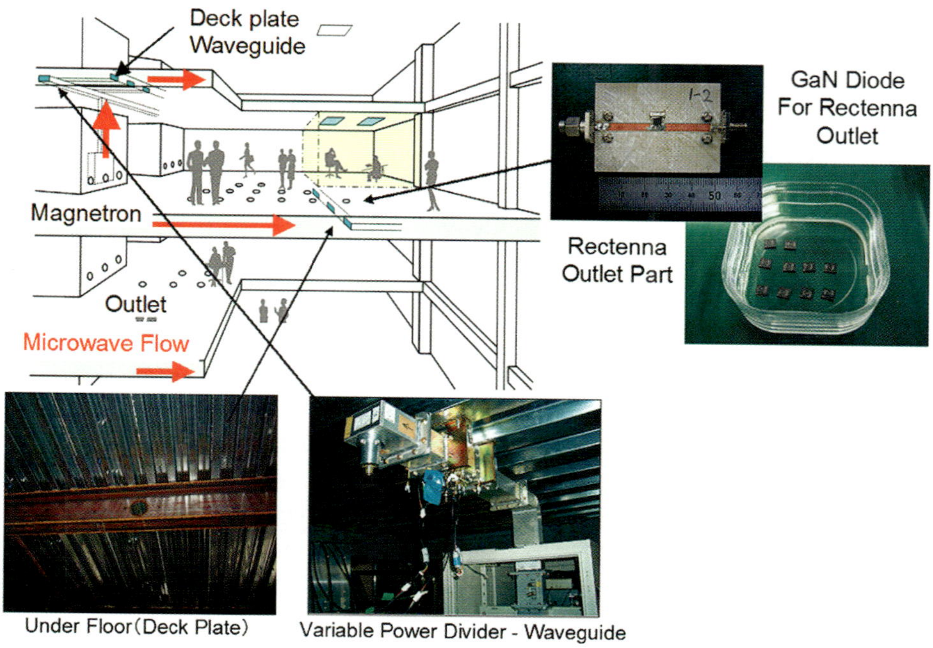


**Figure 9.9** Wireless building using microwave power transmission.

A new GaN Schottky diode has been developed by Tokushima University and Kyoto University in Japan for this application to increase the rectified power and to reduce the charging time. Here, 74.4% efficiency was experimentally achieved at 5 W, 2.45 GHz microwave input with a developed GaN diode whose breakdown voltage is 40 V [9-23]. We also increased the received microwave power with conventional diodes. We have already developed a high-efficiency receiving array called a rectenna with Nihon Dengyo Kosaku Co., Volvo, and Kyoto University, which can rectify 10 kW with over 80% RF-DC conversion efficiency.

The wireless charging system of the EV is not only a short-distance system. Kashima Co., a Japanese building company, jointly proposed with Kyoto University a wireless building using microwave power technology [9-28]. This system supplies electric power wirelessly using building materials (deck plate) that act as microwave transmission waveguides (Figure 9.9). Therefore, the initial cost of the building can be reduced because no wire construction is needed. Microwave power through the deck plate waveguide does not diffuse as it does in free space. Rectennas as dc converters are placed under the floor. The position of the rectenna is easily changed because the microwaves are present practically everywhere under the floor. Therefore, the lifecycle cost of the building can be reduced by converting electricity from dc to microwave to dc. In the initial phase, the wireless system was considered for office buildings where dc-driven computers and other office automation instruments are mainly used. It is estimated that one dc converter requires <50 W, and > 3 kW of microwave power is provided to one room, indicating sufficient power to run a number of typical electrical devices per room. A frequency of 2.45 GHz was selected on the basis of size limitations of a conventional deck plate, and a magnetron was used as the microwave transmitter to reduce the cost.

## 9.5 Conclusions

MPT, a long-distance power transfer system, is one of the candidates for WPT technologies. Compared to the short-distance power transfer characteristics of inductive and resonance coupling, commercial MPT systems have not gained favor yet because of problems with frequency regulation and some technical user requirements that have not been researched yet. Several consortiums, forums, and alliances of short-distance power transfer have been established. They include the Wireless Power Consortium (WPC) for inductive coupling, the Alliance for Wireless Power (A4WP) for resonance coupling, the Power Matters Alliance (PMA) in IEEE-SA for inductive coupling, the Broadband Wireless Forum (BWF) in Japan for both inductive coupling and WPT, the Wireless Power Management Consortium (WP+Mc) in Japan for direct current resonance coupling, and the Korean Wireless Power Forum (KWPF), among others. These have been formed to address the expectations of users. However, to solve technical problems and establish new radio wave regulations, some consortiums have been established in Japan. One is the Wireless Power Transfer Consortium for Practical Applications (WiPoT) for MPT, and the other is the Energy Harvesting Consortium (EHC) for all energy harvesting including EH from broadcasting radio waves.

At the end of the 19th century, no one believed in the feasibility of wireless power. Now, it is generally accepted as a reality. There have been successful WPT and MPT experiments and commercial products. Today, we have a dream that no wire or batteries will be needed for electric power in the near future.

## References


9-1. N. Tesla, "The transmission of electric energy without wires," The Thirteenth Anniversary Number of the Electrical World and Engineer, March 5, 1904.

9-2. N. Tesla, *Experiments with Alternate Current of High Potential and High Frequency*, McGraw Pub. Co., N.Y., 1904.

9-3. W. C. Brown, "The history of power transmission by radio waves," *IEEE Trans. MTT*, MTT-32, no. 9, pp. 1230–1242, 1984.

9-4. W. C. Brown, "The History of the Development of the Rectenna," Proc. of SPS microwave systems workshop at JSC-NASA, pp. 271–280, 1980.

9-5. W. C. Brown, "Adapting Microwave Techniques to Help Solve Future Energy Problems," 1973 G-MTT International Microwave Symposium Digest of Technical Papers 73.1, 1973, pp. 189–191, 1973.

9-6. R. M. Dickinson, "Performance of a high-power, 2.388-GHz receiving array in wireless power transmission over 1.54 km," 1976 MTT-S Int. Microwave Symp. Digest, pp. 139–141, 1976.

9-7. P. E. Glaser, "Power from the Sun; Its Future," *Science*, no. 162, pp. 857–886, 1968.

9-8. H. Matsumoto, "Research on Solar Power Station and Microwave Power Transmission in Japan: Review and Perspectives," *IEEE Microwave Magazine*, pp. 36–45, December 2002.

9-9. Y. Fuse, T. Saito, S. Mihara, K. IJichi, K. Namura, Y. Honma, T. Sasaki, Y. Ozawa, E. Fujiwara, and T. Fujiwara, "Microwave Energy Transmission Program for SSPS," Proc. of International Union of Radio Science (URSI) General Assembly 2011, Proceeding CD-ROM CHGBDJK-2.pdf, 2011.

9-10. S. Sasaki, K. Tanaka, S. Kawasaki, N. Shinohara, K. Higuchi, N. Okuizumi, K. Senda, K. Ishimura, and USEF SSPS Study Team, "Conceptual Study of SSPS Demonstration Experiment," *The Radio Science Bulletin*, no. 310, pp. 9–14, 2004.

9-11. H. Matsumoto, "Research on Solar Power Station and Microwave Power Transmission in Japan: Review and Perspectives," *IEEE Microwave Magazine*, pp. 36–45, December 2002.

9-12. H. Matsumoto, N. Kaya, M. Fujita, Y. Fujino, T. Fujiwara, and T. Sato, "MILAX Airplane Experiment and Model Airplane (in Japanese)," Proc. of the 11th ISAS Space Energy Symposium, pp. 47–52, 1993.

9-13. Y. Fujino, T. Itoh, M. Fujita, N. Kaya, H. Matsumoto, K. Kawabata, H. Sawada, and T. Onodera, "A Rectenna for MILAX," Proc. of Wireless Power Transmission Conference '93, pp. 273–277, 1993.

9-14. T. Mitani, H. Yamakawa, N. Shinohara, K. Hashimoto, S. Kawasaki, F. Takahashi, H. Yonekura, T. Hirano, T. Fujiwara, K. Nagano, H. Ueda, and M. Ando, "Demonstration Experiment of Microwave Power and Information Transmission from an Airship," Proc. of 2nd International Symposium on Radio System and Space Plasma 2010, pp. 157–160, 2010.

9-15. N. Shinohara, "Beam Control Technologies with a High-Efficiency Phased Array for Microwave Power Transmission in Japan," Proceedings of IEEE, vol. 101, Issue 6, pp. 1448–1463, 2013.

9-16. N. Shinohara, H. Matsumoto, and K. Hashimoto, "Phase-Controlled Magnetron Development for SPORTS: Space Power Radio Transmission System," The Radio Science Bulletin, No. 310, pp. 29–35, 2004.

9-17. R. J. Gutmann and R. B. Gworek, "Yagi-Uda Receiving Elements in Microwave Power Transmission System Rectennas," *Journal of Microwave Power*, vol. 14, no. 4, pp. 313–320, 1979.

9-18. T. W. Yoo and K. Chang, "Theoretical and Experimental Development of 10 and 35 GHz Rectennas," IEEE Trans. MTT, vol. 40, no. 6, pp. 1259–1266, 1992.

9.19. N. Shinohara, T. Mitani, and H. Matsumoto, "Study on Ubiquitous Power Source with Microwave Power Transmission," Proc. of International Union of Radio Science (URSI) General Assembly 2005, Proceeding CD-ROM C07.5(01145).pdf, 2005.

9-20. K. Hatano, N. Shinohara, T. Mitani, K. Nishikawa, T. Seki, and K. Hiraga, "Development of Class-F Load Rectennas," Proc. of 2011 IEEE MTT-S International Microwave Workshop Series on Innovative Wireless Power Transmission: Technologies, Systems, and Applications (IMWS-IWPT2011), 2011.

9-21. K. Hatano, N. Shinohara, T. Seki, and M. Kawashima, "Development of MMIC Rectenna at 24GHz," Proc. of 2013 IEEE Radio & Wireless Symposium (RWS), pp. 199–201, 2013.

9.22. T. Ichihara, T. Mitani, and N. Shinohara, "Study on Intermittent Microwave Power Transmission to a ZigBee Device," Proc. of 2012 IEEE MTT-S International Microwave Workshop Series on Innovative Wireless Power Transmission: Technologies, Systems, and Applications (IMWS-IWPT2012), pp. 209–212, 2012.

9-23. N. Shinohara, "Wireless Charging System of Electric Vehicle with GaN Schottky Diodes," IMS2011 Workshop WFA "Wireless Power Transmission," 2011.

9-24. A. P. Sample and J.R. Smith, "Experimental Results with Two Wireless Power Transfer Systems," Proc. of 2009 IEEE Radio & Wireless Symposium (RWS), MO2A-5, pp. 16–18, 2009.

9-25. J. R. Smith, "Mapping the space of wirelessly powered systems," Proc. of IMS 2010 Workshops, WFB-3, 2010.

9-26. N. Shinohara, "Beam Efficiency of Wireless Power Transmission via Radio Waves from Short Range to Long Range," *Journal of the Korean Institute of Electromagnetic Engineering and Science*, vol. 10, no. 4, pp. 224–230, 2011.

9-27. N. Shinohara, Y. Kubo, and H. Tonomura, "Mid-Distance Wireless Power Transmission for Electric Truck via Microwaves," Proc. of 2013 International Symposium on Electromagnetic Theory (EMT-S2013), 2013.

9-28. N. Shinohara, Y. Miyata, T. Mitani, N. Niwa, K. Takagi, K. Hamamoto, S. Ujigawa, J.-P. Ao, Y. Ohno, "New Application of Microwave Power Transmission for Wireless Power Distribution System in Buildings," Proc. of Asia-Pacific Microwave Conference 2008, CD-ROM H2-08.pdf, 2008.

# Chapter 10
## Industrial Applications of WPT

The industrial wireless technology works on the induction principle, similar to a transformer's primary/secondary transfer. In a transformer, the primary and secondary windings are on a common, closed ferromagnetic core. For long tracks, on the other hand, WPT "stretches" the primary winding to a long loop and places the secondary winding onto an open ferromagnetic core, as described schematically in Figure 10.1. This allows relative motion of the two windings. The transmission behavior is optimized by using a high transmission frequency of 20 kHz or 25 kHz.

During the development of the technology, a strong focus is placed on important criteria such as efficiently covering a wide range of performance, multiple areas of application, but also the greatest possible level of environmental compatibility.

A particularly high degree of efficiency as well as extremely good electromagnetic compatibility (EMC) result from an exceptionally low track current of only 70 A (with special applications, even as low as 35 A). In addition to pure power supply, additional system features such as inductive data transfer and inductive track guidance may be integrated into the product portfolios to meet diverse material handling requirements.

## 10.1 System Overview

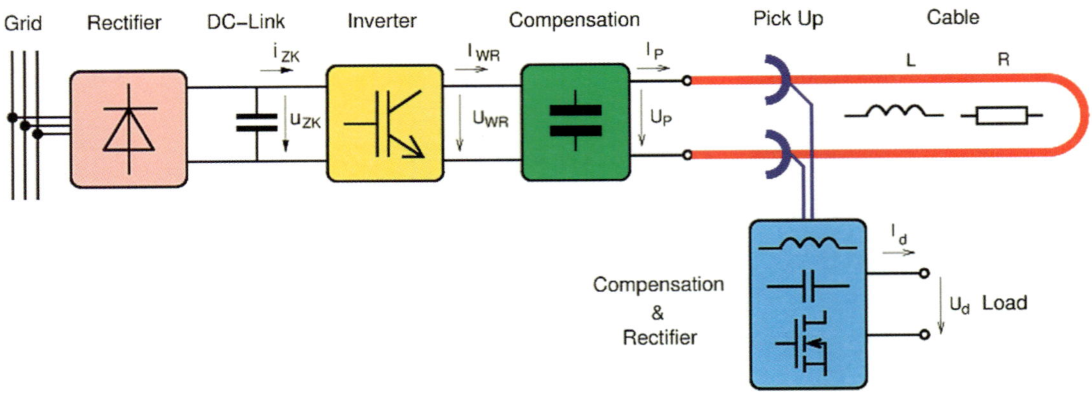


**Figure 10.1** A WPT system overview of an industrial plant.

The primary inverter converts the customer's existing three-phase alternating current into single-phase ac-voltage with the working frequency. Therefore, a passive rectifier with a grid EMI-filter is used to rectify the grid voltage. Afterward, the resulting dc-voltage is converted into an ac squared voltage at the output of the inverter, as shown in Figure 10.2. The conventional full-bridge converter works in a resonant mode, in such a way that switching losses of the IGBTs are minimal. Thus, the working frequency of the inverter is equal to the resonance frequency of the compensation circuit.

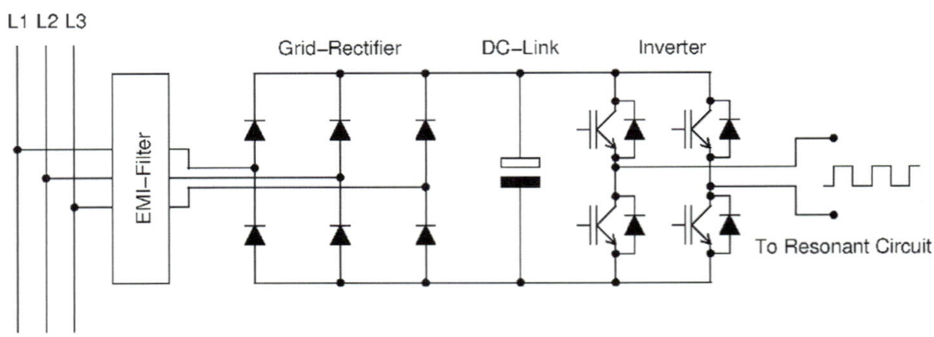


**Figure 10.2** Primary inverter circuit.

Due to the series connection of multiple secondary loads, the constant voltage supply of the primary is possible if only one load is placed on the secondary. In most cases, more than one load is required on the primary track; thus, the system is fed by constant current. Therefore, a gyrator circuit is used to transform the constant ac-voltage out of the inverter into a constant ac-current into the primary coil [10-1]. The behavior of the gyrator circuit was described in detail in chapter 8.

The voltage induced in the pick-up coil is rectified and adapted to the consumer requirements. The simplest circuit is a non-active rectifier, which does not influence the power flow. Two typical examples are shown in Figure 10.3(a) and Figure 10.3(c). The parallel resonant circuit (a) behaves as a current source, and the series resonance circuit (c) behaves as a voltage source, assuming the primary winding is fed by a constant current.

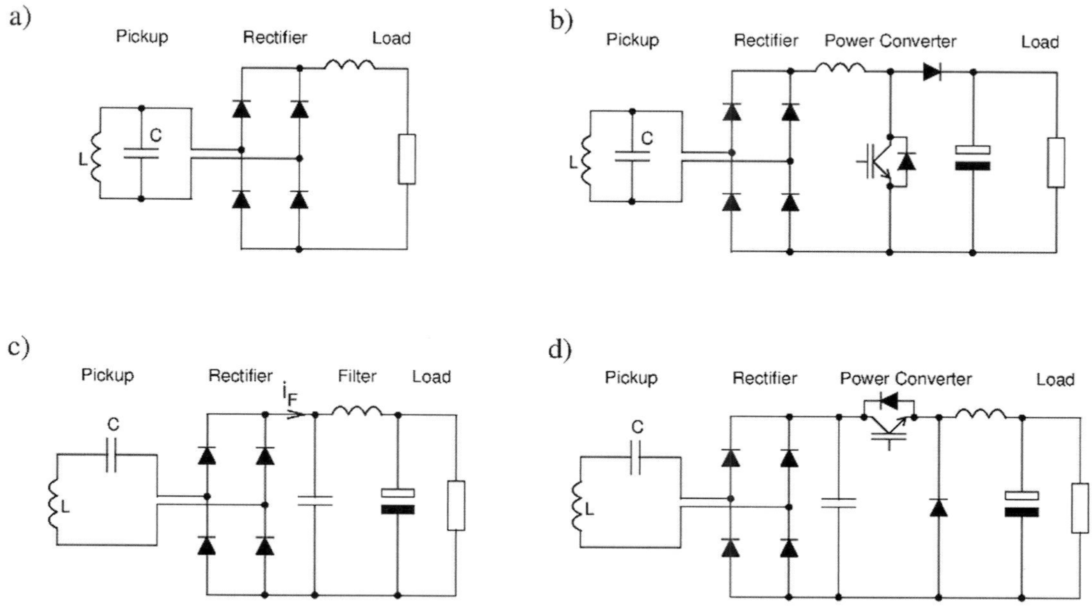


**Figure 10.3** Secondary rectifier topologies.

On the other hand, when the output voltage or current has to be regulated, the secondary power electronics need to actively influence the power fed into the secondary circuit by means of a switched mode power supply. A typical way to realize this is to implement a dc-dc converter into the secondary components. Figure 10.3(b) and Figure 10.3(d) show two possible converter topologies. Most of the industrial application pick-up devices use a series compensation because a constant voltage is a more common interface than a constant current. Figure 10.4 shows the composition of the overall efficiency of the system.

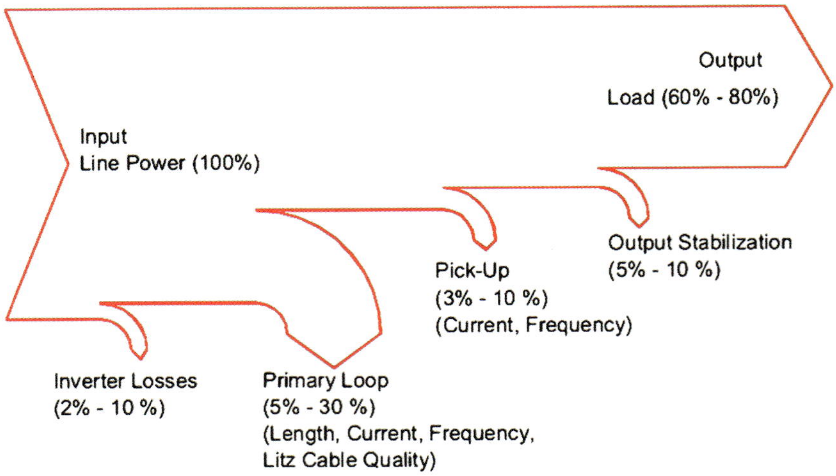


**Figure 10.4** Efficiency distribution on the components and their dependency.

- Assuming that the line power is set to 100%, the losses of the primary inverter are in the range of 2% to 10% depending on the line voltage, the inverter design, and the inverter control concept.
- The losses of the primary loop (5% to 30%) mainly depend on the length, the Litz cable cross section, and the quality.
- The transfer losses mainly depend on the pick-up design and can be calculated in the range of 3% to 10%. Because of the useful magnetic coupling to the primary loop of the E-shaped pick-up, this design is characterized by high transfer efficiency.
- Depending on the pick-up compensation concept, it might be necessary to add an active stabilization unit to the system. Additional losses of 5% to 10% have to be taken into consideration. In summary, the overall efficiency of an inductive power supply system is about 60% to 80%. [10-2].

Figure 10.5 shows the known configurations for an industrial application of a straight line type. The flat pick-ups may be fed by a cable laid with a guide profile (a) or directly from the floor (b). For E- or U-shaped pickups, the cable may be mounted on extended supports (c) or in an electric monorail system (EMS) profile (d).

These wireless systems feature absolutely maintenance-free operation at the highest level due to the lack of friction and contact. A wide power range is provided by scalable hardware under the best EMC/EMF behavior.

The delivered systems may be combined with an integrated data transfer to forward control data in a bidirectional way by modulating the data signals on the primary wires. Additionally, a track guidance system may be implemented to give information on the relative lateral position of the vehicle to the primary track by sensing the primary magnetic field [10-3].

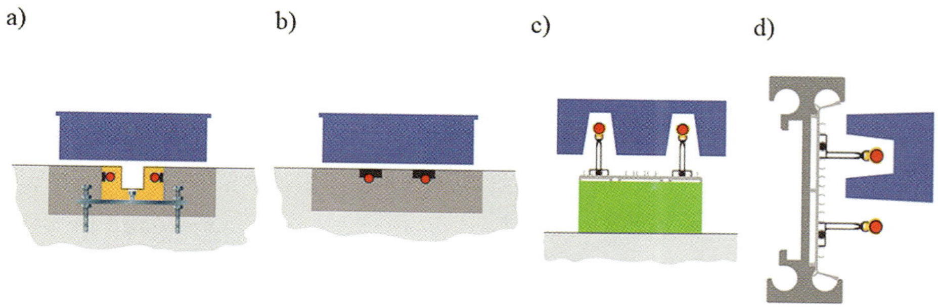


**Figure 10.5** Different application examples of power supply cable installations and onboard pick-ups for industrial wireless power transfer.

## 10.2 Crane Applications

One of the first applications of wireless power was done in the United States in 1999 for a harbor crane in Virginia, as shown in Figure 10.6. Even this early application showed the robustness of wireless power transfer with a rated power of about 40 kW. The used inverter had an active controlled current output, and the secondary circuit was built out of two E-shaped pickups. The system worked stably under harsh environmental conditions. In 2004, the crane was crashed into by a ship and was no longer operable. Today, such applications are widespread and are one of the standard solutions for crane powering.

**Figure 10.6** Crane 2 × 22 kW equipment in Virginia, USA (courtesy of Paul Vahle GmbH & Co.KG).

## 10.3 Automated Guided Vehicles

**Figure 10.7** Left: AGV 3 kW Super-Capacitor chargers in Coburg (Kaeser), Right: 1.5 kW AGV-supply with track guiding sensor in Tubos del Mas, Spain.

In the past, mainly batteries or underground conductor lines were used to supply the electrical power of an automated guided vehicle (AGV). Today, an increasing number of contactless inductive power supply units are being deployed, as shown in Figure 10.7. In connection with a track guidance, which is also inductive, the user can benefit from a completely smooth floor surface.

This increases plant availability substantially, as the system is nearly insensitive to dirt, oil, and other types of contaminants. Thanks to the vast variety of available pickup units, an inductive supply to automated guided vehicles of virtually any size category is possible.

A main advantage of inductive power supply is a barrier-free track path with the possibility of complex track layouts. The system can easily be expanded and is able to charge a battery or ultracapacitor during travel.

## 10.4 Skillet Conveyor

Conventional power supplies (conductor lines) underneath skillet platforms require high maintenance because they are sensitive to contaminants, whereas no maintenance is required when using wireless technology. Mechanical damage of power supply components is generally prevented, even in critical track sections such as at lifting stations and transfer stations.

As depicted in Figure 10.8, a flat pick-up system or an E-shaped one are available for this application. The flat pick-ups allow a high degree of movement flexibility. In E-shaped pick-up systems for skillet conveyors, the coil encompasses the primary cable. Owing to the particularly efficient electromagnetic coupling, high performance is provided, even with small pick-up construction sizes.

**Figure 10.8** Automotive application in Neckars Ulm (Audi): 3kW flat pickup and 4kW E-shaped pickup for automotive production (Vahle).

The advantages of this solution against conventional techniques are the trouble-free and safe "threading" on transfer stations and its maintenance-free operation, insensitivity to dirt, and high availability.

## 10.5 Transfer Car

In heavy industry (e.g., steel plants or aluminum plants), general warehouse, or even clean-room applications, rail-mounted transfer cars are frequently used for transporting material between the different working stations or storage facilities, as shown in Figure 10.9. If wireless technology is used for supplying power, the track is completely free of disturbing installations (conductor lines, cable reel, and so on) and is traversable without obstruction for possible cross traffic. The transfer car's application is mostly based on flat pick-up systems.

**Figure 10.9** Metal industry 1.5 kW equipment in Alicante (Alcoa).

## 10.6 Electric Monorail System

In the automotive industry, as well as other industries, an electric monorail system (EMS) is used as a rail-bound conveying means with numerous individually driven vehicles (carriers). At the same time, the runway is used as a support profile and carries the wireless power transfer components necessary for the power supply and control supply to save space. Branches are created using track switches. Lifters are used for transporting the payload onto other manufacturing levels or for coupling different production steps. Using wireless power transfer increases plant availability substantially because no abrasive wear occurs, thereby making the system maintenance-free.

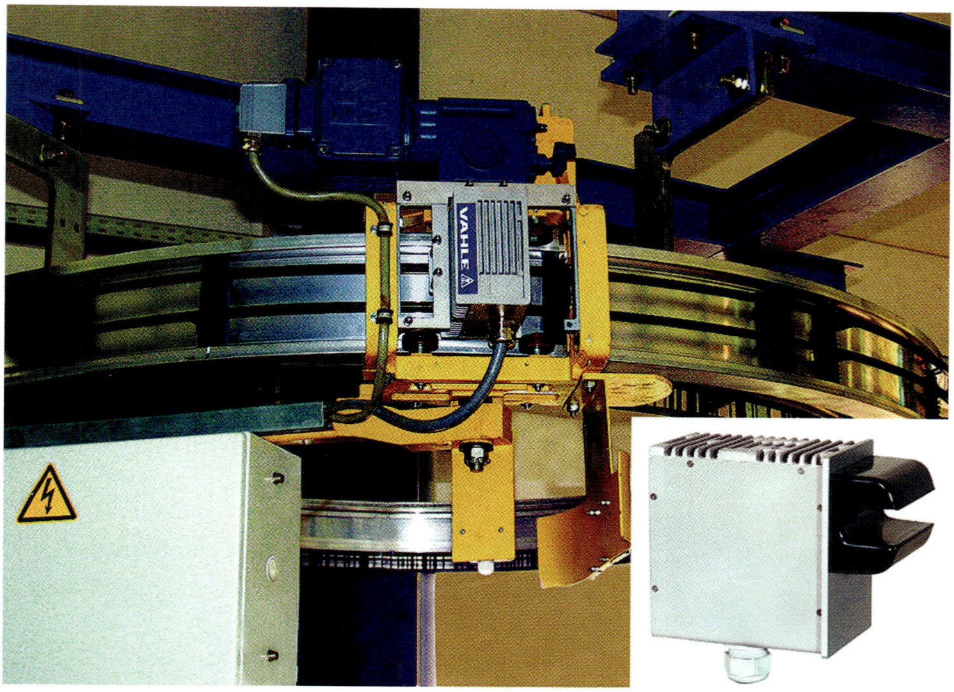


**Figure 10.10** Test plant with 1.5 kW U-shaped pick-up in Ditzingen (Siemens).

U-shaped pick-up systems, as shown in Figure 10.10, are mounted on a twin conductor system specifically developed for this application. It has proven its suitability thanks to its excellent EMF and EMC behavior.

One of the important features is the "C1"-standard conformity of the system [10-4], which allows it to fit most EMS Systems. Thus, it is applicable for complex track layouts with no contamination of conveyed material due to carbon abrasion, as is the case with sliding contacts. It is easy to install, due to special fixture technology, and has no restriction to travel speed.

## 10.7 Sorter Technology

For transporting and sorting luggage items, packages, and other general cargo, distribution centers use large sorting equipment that presorts the arriving cargo and prepares it for further transport to different target destinations.

Power supply components must meet high demands with respect to exceedingly narrow space conditions as well as very high travel speeds usually occurring in the sorting technology. There is a range of I-shaped pick-ups that can be installed in a compact manner to fulfill these requirements. Figure 10.11 shows one common pick-up for this application.

**Figure 10.11** A 150 W pick-up in Mayoral (Sandvik/Cinetic).

Even at the highest travel speed, no noise is generated by the pick-up due to its mechanical design.

## 10.8 Clean-Room Technology

Special operating conditions, such as high travel speeds, extremely difficult environmental influences, but also high demands for a clean production environment (clean room) require a special form of power supply. Wireless power transfer is the optimal solution to meet these requirements. Meanwhile, this technology is very frequently used for stacker cranes or overhead transport conveyors in clean-room technology, for example, as shown in Figure 10.12.

**Figure 10.12** E-shaped pick-ups (2 × 22 kW) for TFT-display production/clean-room (Sharp/Japan).

Due to the high electrical drive capacity often used in clean-room technology, the use of U- or E-shaped pick-ups is a suitable option. These designs, in particular, have an extremely good coupling quality to the track conductor's magnetic field and, therefore, enable optimal power transfer.

Unlike sliding contact technology, there is no contamination of the environment caused by carbon abrasion. The systems should be classified in the clean-room class 1 in accordance with U.S. Fed. Std. 209 and ISO 14644 [10-5].

## 10.9 Elevator Systems

As an alternative to the traveling cable, wireless power transfer perfectly meets the requirements when power must be supplied to an elevator cab in a maintenance-free and fail-safe manner, regardless of the ambient conditions. Whether it is for the standard elevator or the inclinator, this alternative offers elevator systems completely new opportunities without any restrictions to speed or elevation heights.

**Figure 10.13** Elevator application and 5 kW flat pick-up (Cinema in Moskau).

Depending on the required capacity and existing installation space at the elevator cab, both flat pick-ups and U-shaped pick-ups are suitable for this application.

## References

10-1. J. Meins, F. Turki, and R. Czainski, "Contactless High Power Supply," Proc. of UEES '04 Conference, pp. 581–586, Alushta, 2004.

10-2 J. Meins, G. Buehler, R. Czainski, and F. Turki, Contactless Inductive Power Supply. Maglev Conference 2006, pp. 527–535, Dresden, 2006.

10-3. Paul Vahle GmbH & Co.KG, http://www.vahle.de, Accessed February 2015.

10-4. VDI-Guideline 3643: Elektrohängebahn, Obenläufer, Traglastbereich 500 Kg (1998–11).

10-5. International Standard ISO 14644, Cleanrooms and controlled environments—Part 1: Classification of air cleanliness, and Part 2: Specifications for testing and monitoring.

# Concluding Remarks

The title *Wireless Charging Technology and the Future of Electric Transportation* elicits notions of convenience, autonomy, and efficiency in all aspects of application, whether consumer electronics or industrial automation and on into the various modes of transportation, be they personal mobility devices or automobiles, trains, planes, and ships. Regardless of the application or mode of electrified transportation, wireless power transfer (WPT) not only has emerged as the enabling technology that will foster deeper penetration of electrification into the marketplace, but also it shows the potential to open new avenues in convenience and autonomy that could only be dreamed of just a few short years ago.

This text has presented the views of WPT experts from around the globe who have put their thoughts and experiences on paper to share where this technology is now and the potential it has to change the future, particularly of transportation. In doing so, it will accelerate an already existing paradigm shift to green transportation and renewable energy.

# Index

# About the Authors

## In-Soo Suh, Lead Author

**In-Soo Suh** is an associate professor in the Cho Chun Shik Graduate School for Green Transportation, KAIST. His research fields are electric vehicle systems with wireless and conductive charging infrastructure strategies, vehicle system integration focused on the electrical powertrain, noise vibration harshness (NVH) and structural acoustics, and future intelligent transportation technology related to smart grids and communication technology. In addition to Dr. Suh's 15-year industrial experiences in global automotive OEMs with Chrysler Corporation and GM Korea, he pursues the expansion of his career in academia with emphasis on applied engineering research and education. As a vehicle group manager in KAIST's Wireless Power Transfer Project Group, during 2009-10, an important contribution was to lead the systematic efforts in the historic launch of the OLEV system in Seoul Grand Park and to build ten different types of OLEV vehicles. In the fall of 2013, he invented and unveiled a foldable microelectric vehicle, called "Armadillo-T," a prototype concept vehicle for future urban application of micro mobility. He is an associate editor of IEEE Intelligent Transportation System Magazine and a research planner of Korean National Research Foundation. Dr. Suh holds a PhD in mechanical engineering from Massachusetts Institute of Technology, specializing in structural acoustics. He obtained a BS and MS in mechanical engineering from Seoul National University, Korea.

## Dong-Ho Cho

**Dr. Dong-Ho Cho** has been a professor in the Department of Electrical Engineering of KAIST since 1998, and he was a director of KAIST Institute for Information Technology Convergence from 2007 to 2011. He has been a director of KAIST Online Electric Vehicle Project since 2009, and he has served as the head of The Cho Chun Sik Graduate School for Green Transportation since 2010. He was a vice president of KAIST from 2011 to 2013. From 1987 to 1997, Dr. Cho was a professor in the Department of Computer Engineering at Kyunghee University. He received his PhD in electrical engineering from KAIST. His major research interests include mobile communication, the online electric vehicle system based on wireless power transfer, and genomics based on bioinformatics.

## Jörg Franke

**Prof. Dr.-Ing. Jörg Franke** has headed the Institute for Factory Automation and Production Systems (FAPS) at the Friedrich-Alexander-University of Erlangen-Nuremberg, Germany, since 2009. His research focuses on manufacturing of mechatronic products, starting from packaging of electronic circuits, structured metallization of circuit carriers, surface mount technology, assembly of electric motors, and automation solutions; and ending with engineering, planning, and simulation of complex mechatronic systems. Previously, he led management positions with global responsibilities at McKinsey&Co, Robert Bosch GmbH, ZF Lenksysteme GmbH, Schaeffler AG, and ABM Greiffenberger AG. Jörg Franke studied and prepared his doctor's thesis in the field of production engineering.

## Dr. Soon-Man Hong

**Dr. Soon-Man Hong** has been a visiting professor of the Graduate school for Green Transportation at KAIST, Korea, from 2014, since his completed term of president of Korea Railroad Research Institute (KRRI), Korea. He has been actively involved in the transportation sector with the Ministry of Land, Transport and Maritime Affairs (MLTM), Korea. He served as the Deputy Minister for the Office of Transport Policy in 2010, the Assistant Minister as the Head of Aviation Safety Authority, and the Director General for Railroad Policy in MLTM, Korea. He was also a visiting professor of the Graduate School for Green Transportation at KAIST just before he joined KRRI. Dr. Hong obtained a PhD and an MS in Civil Engineering (majored in transportation) from the University of Washington, and he also holds a CPA license.

## Sung-Kwan Jung

**Sung-Kwan Jung** has been a research associate professor of IT Convergence (KIITC) at KAIST Institute since 2012. From 2007 to 2012, he was a senior researcher at KIITC. At KIITC, he has researched wireless power transfer technologies for consumer devices, especially on small mobile devices. He also researched wireless power transfer and communication technologies for small sensor nodes. For the last three years, he studied resonant wireless power transfer technologies for small mobile devices in home and office application environments. Currently, his research interest is the wireless power-data co-transfer technology for lightweight sensor nodes including M2M (machine-to-machine) and IoT (Internet-Of-Things) devices. As a team leader in KIITC, he also has research interests in IoT/WoT (Web-Of-Things) protocol and architecture, sensor-web integration, and UI/UX (user interface/user experience) for smart devices. Dr. Sung-Kwan Jung received a BS, MS, and PhD in electrical engineering from the Korea Advanced Institute of Science and Technology (KAIST), Korea.

## Byung-Song Lee

**Dr. Byung-Song Lee** has been a senior and principal researcher at the Korea Railroad Research Institute since 1998. He is a leader of the Railroad Electric Traction & Wireless Power Transfer System Research Team. His research interests include wireless power transfer and energy conversion systems for railway trains. Currently, he is working on development of wireless power transfer systems and a linear pulse motor for railway vehicles. From 1996 to 1997, he worked as a senior researcher in the Railway Vehicle R&D Center at KHRC (Korea High Speed Rail Construction Authority), where he developed the propulsion system for high-speed trains. He received a BS in electrical engineering from Seoul National University of Technology, and an MS and PhD in electrical engineering from Chung-Ang University, Korea.

## John M. Miller

**John M. Miller** was a distinguished research scientist at the Oak Ridge National Laboratory (ORNL) and former director of the Power Electronics and Electric Power Systems Research Center prior to his retirement in 2014. Dr. Miller's research interests include all aspects of wireless power charging of electric vehicles, including advancement of international standards on wireless power transfer, interoperability, and control of leakage fields. He is also active in electric machine design, specifically electric traction motors for plug-in and battery electric vehicles (PEV). In the area of PEV traction drive systems, he is an advocate of motor rating standardization, performance metrics, and efficiency standardization. He received a BSEE from the University of Arkansas, MSEE from Southern Methodist University, and PhD in electrical engineering from Michigan State University.

## Florian Risch

**Florian Risch** has worked at the Institute for Factory Automation and Production Systems (FAPS) of Friedrich-Alexander-University of Erlangen-Nuremberg, Germany, since 2008. Since 2011 he has served as a team leader of the research group of the Bavarian Technology Center for Electric Drives (E|Drive-Center). His research includes stationary and in-motion wireless power transfer systems for electric vehicles, with a focus on planning and verifying different fields of application and the development of efficient production processes for wireless charging infrastructure and vehicle components. Mr. Risch received a Dipl.-Wirtsch.-Ing. from Friedrich-Alexander-University of Erlangen-Nuremberg, Germany.

## Naoki Shinohara

**Naoki Shinohara** has been a professor at the Research Institute for Sustainable Humanosphere, Kyoto University, Japan, since 2010. He has been engaged in research on the Solar Power Station/Satellite and Microwave Power Transmission System. He was a research associate at the Radio Atmospheric Science Center, Kyoto University from 1996 to 2000. He was a research associate at the Radio Science Center for Space and Atmosphere, Kyoto University from 2000 to 2001, and there he was an associate professor from 2001 to 2004. He was an associate professor at the Research Institute for Sustainable Humanosphere, Kyoto University from 2004 to 2010. Prof. Shinohara received BS, ME, and PhD degrees in electrical engineering from Kyoto University, Japan.

## Faical Turki

**Dr. Faical Turki** is a power electronics specialist with more than 14 years' experience in research and development of industrial electronics. He spent four years researching IGBT-based power electronics for wind turbines, and was involved in development of the first wireless inductive power supply systems. Starting with his work in the Institute for Electrical Machines, Traction and Drives in Braunschweig, Germany, he holds more than ten years of research experiences in the field of wireless power supplies. Since 2007, he has worked at Paul Vahle GmbH & Co.KG, Germany, in the development department for Contactless Power Systems CPS®, and he is currently head of the research and development of contactless power and data transfer systems for electric vehicles, trams, and industrial applications. His research includes polyphased inductive high power supply, bidirectional approaches, magnetic design, high-frequency resonant converter topologies, and data transfer on magnetically coupled power transfer circuits. Dr. Turki received the Dipl.-Ing. degree in electrical engineering and a PhD in electrical engineering at the Technical University of Braunschweig, Germany.